PRINCIPLES
OF
ILLUMINATION

by

John E. Traister

HOWARD W. SAMS & CO., INC.
THE BOBBS-MERRILL CO., INC.
INDIANAPOLIS · KANSAS CITY · NEW YORK

FIRST EDITION

FIRST PRINTING—1974

Copyright © 1974 by Howard W. Sams & Co., Inc., Indianapolis, Indiana 46268. Printed in the United States of America.

All rights reserved. Reproduction or use, without express permission, of editorial or pictorial content, in any manner, is prohibited. No patent liability is assumed with respect to the use of the information contained herein. While every precaution has been taken in the preparation of this book, the publisher assumes no responsibility for errors or omissions. Neither is any liability assumed for damages resulting from the use of the information contained herein.

International Standard Book Number: 0-672-20973-X
Library of Congress Catalog Card Number: 73-83361

Preface

The subject of illumination has been covered by many technical books, mostly from the standpoint of a formal presentation of basic theory. What would appear to be helpful to most people is practical information on basic theory, extended in a way to facilitate a smooth transition from the book to actual applications. This book is designed to provide information necessary for this transition.

It is not the purpose of this book to cover the entire field of illumination, but rather a generous sampling of the many features of modern practice. Units are planned to provide instructional material for students, craftsmen, technicians, contractors, and others in the illumination field who need to obtain the basic fundamentals of illumination.

Section I is a general introduction to illumination. Section II covers light sources like incandescent lamps, mercury lamps, and fluorescent lamps. The principles of lighting design are covered in the next section. Section IV discusses interior lighting design of offices, schools, stores, residences, and industrial buildings. Exterior lighting of sports fields, underwater areas, roadways, and signs are discussed in Section V. Next, the special lighting applications of germicidal lamps, sun lamps, and lamps for use in horticulture are discussed. Section VII deals with lighting cost analysis, and Section VIII discusses lighting controls and wiring. Also, a generous coverage of recommended illumination levels and luminaire coefficient tables are given in the appendixes.

I wish to thank C. Keeler Chapman whose illustrations were invaluable to me. I also wish to express my appreciation for the many hours of typing by Mrs. Ruby Updike whose willing assistance has made the preparation of this book much easier than it otherwise would have been.

JOHN E. TRAISTER

Contents

SECTION I—INTRODUCTION

Unit   *Page*
1. Introduction to Illumination 9
2. The Eye and Vision 13
3. Characteristics and Measurement of Light 17

SECTION II—LIGHT SOURCES

4. Lamp Classifications 23
5. Incandescent Lamps 27
6. Mercury Lamps 35
7. Fluorescent Lamps 39

SECTION III—PRINCIPLES OF LIGHTING DESIGN

8. Basic Design Procedure 45
9. Lumen Method Zonal-Cavity System 49
10. Point-by-Point Method 57

SECTION IV—INTERIOR LIGHTING DESIGN

11. Offices and Schools 63
12. Store Lighting 69
13. Residential Lighting 79
14. Industrial Lighting 83

SECTION V—EXTERIOR AND SPORTS LIGHTING

15. General Floodlighting Design 93
16. Sports Lighting 99
17. Underwater Lighting 107
18. Roadway Lighting 111
19. Sign Lighting 121

SECTION VI—SPECIAL LIGHTING APPLICATIONS

20. Germicidal Lamps 127
21. Sunlamps 129
22. Lamps for Use in Horticulture 133
23. Miscellaneous Applications 135

SECTION VII—LIGHTING COST ANALYSIS

24. The Cost of Lighting 141

SECTION VIII—LIGHTING CONTROLS AND WIRING

25. The Control of Lighting 147

SECTION IX—APPENDIX

A. Recommended Illumination Levels 157
B. Coefficient of Utilization Tables 165

Index 171

SECTION I

INTRODUCTION

UNIT 1

Introduction to Illumination

The sun is our greatest source of energy, including light, and man has depended upon it for thousands of years to provide light for work and play. However, not being satisfied with the daylight hours given them by the sun, men have tried numerous ways to create light, in order to see during the hours of darkness and thus make better use of this time. The sketches (as shown in Fig. 1-1) show early attempts at manufacturing this artificial light.

Probably the first artificial lights were burning wooden torches carried about in the hands. Then man attempted to improve this light by selecting wood which burned the brightest and longest; later, animal and vegetable oils were burned in shallow stone dishes. Some lamps of this type have been found which are at least 5000 years old.

The lamps became more ornamental and were made of pottery and metals. Other than the addition of a wick, little improvement was made in the amount of light given off by these lamps for several hundred years.

Wax and tallow candles were also a popular form of light for many years. Candle holders with large numbers of candles in them were used to give a greater source of light for large rooms and auditoriums.

Still, all of these sources of light were inclined to flicker, give off smoke, and were inconvenient.

THE APPLICATION OF ELECTRICITY FOR LIGHT

Electric arcs or flames drawn between two carbon electrodes were one of the first types of electric light. Although the first light of this type was invented about 1801, it was not used commercially until one was installed in the Dungeness lighthouse in England in 1862. While this light was not entirely steady or free from smoke, it was able to produce a great amount of bright blue light as current leaped and spluttered between the two carbon rods. Another drawback with this type of light was the carbon rods which had to be replaced every few hours.

About 1840, many experiments began with electrical incandescent lamps. One such experiment was the heating of high resistance metal or carbon strips to a glowing temperature by passing electric current through them. But none of these were successful or practical until Thomas A. Edison (Fig. 1-2) invented the carbon filament incandescent lamp in 1879.

In a dusty laboratory at Menlo Park, New Jersey, Edison applied electrical current to very thin filaments of carbonized thread sealed in a glass bulb from which the air was removed by vacuum pumps. The voltage was gradually increased until the incandescence built up to a steady bright glow. Electric current kept the filament glowing for 40 hours before Edison deliberately increased the voltage and burned out the lamp.

Edison announced to a small, enthusiastic group that if it could burn for 40 hours, he would make it burn for 100 hours. This moment (October 21st, 1879) marked the beginning of a new era for electric lighting.

Edison also developed the first efficient electric generators to supply current for his lamps, and in 1882 Edison's Pearl Street station started serving 59 customers by providing power to 1284 lamps in the Wall Street section of New York City.

THE ELECTRIC LAMP TODAY

Changes in the electric lamp have kept pace with other changes in the industry during the

PRINCIPLES OF ILLUMINATION

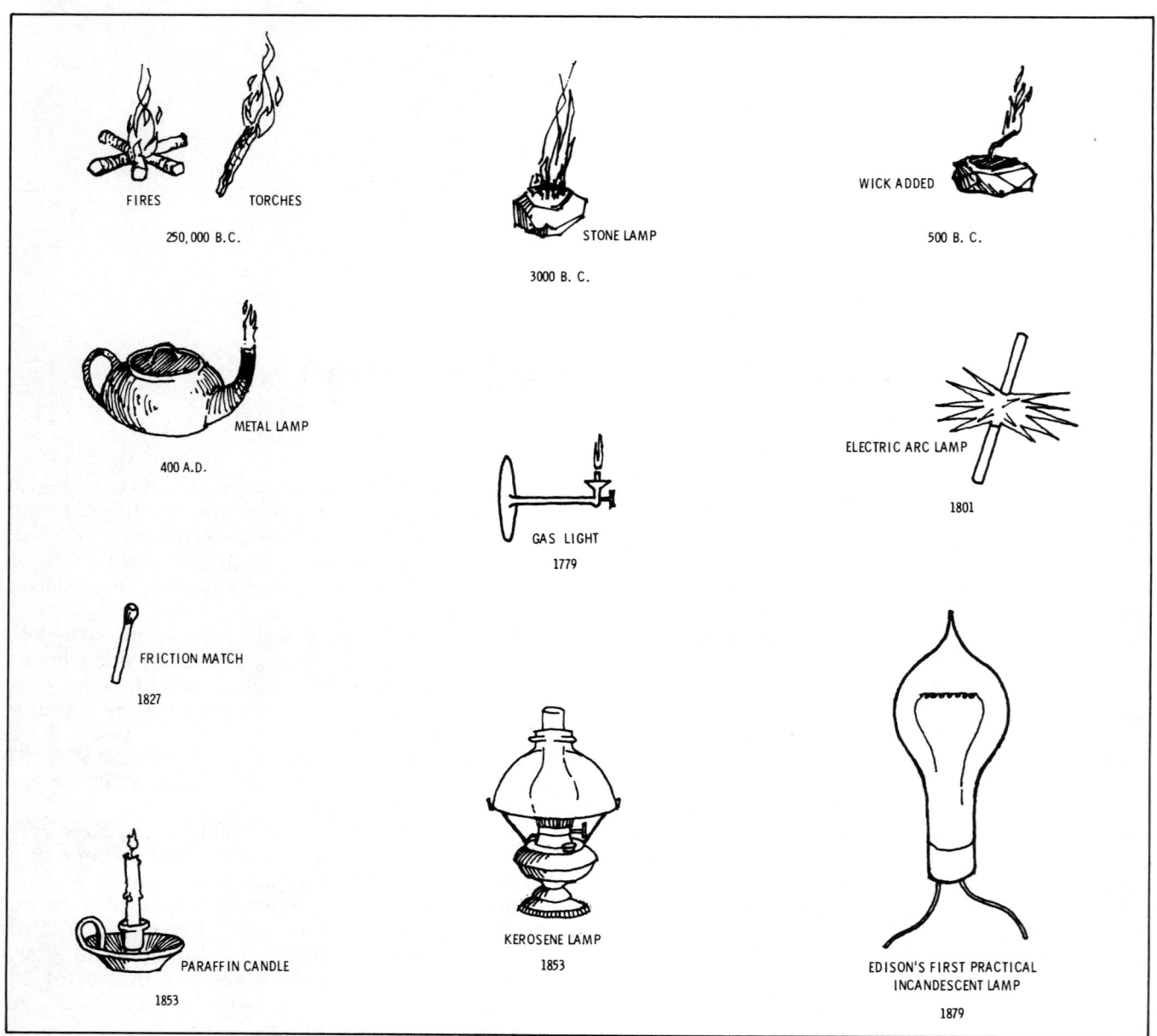

Fig. 1-1. Sketches of early attempts to manufacture artificial light.

years that followed Edison's inventions. His original incandescent lamp had a life of only 40 hours (Fig. 1-3); today, some incandescent lamps have a life of 6000 to 8000 hours (Fig. 1-4). Lamps in Edison's time were manufactured at the approximate rate of 150 per day. Now, machines are capable of producing finished lamps at the rate of 72,000 an hour.

Today, modern electric illumination is one of our greatest advantages, and the business of generating, transmitting, and distributing electricity is now the largest industry in the country—having a total investment of over $70 billion.

CONTRIBUTION TO TWENTIETH-CENTURY PROGRESS

Electric light in the home greatly improves the appearance of the home as well as people and objects in the home. It also speeds up household chores, reduces eye strain, and makes it a pleasure for members of the family to work or play during evening hours. Electric lighting is not only cleaner, safer, and more convenient than any other form of artificial light, it is also low-cost enough to be within the means of almost every family.

INTRODUCTION TO ILLUMINATION

Courtesy General Electric Co.
Fig. 1-2. Picture of Thomas A. Edison.

Courtesy General Electric Co.
Fig. 1-3. A replica of Edison's first lamp, with a life of 40 hours.

In industry, electric lighting speeds up production, reduces errors, increases safety, and generally improves the morale of employees.

In stores, hotels, and office buildings, electric illumination is used on a large scale to improve the efficiency of employees, to aid in the selling of merchandise, and to reduce eye strain.

The exteriors of buildings are beautifully floodlighted and streets are lighted brightly with electric lamps. The lighting of outdoor sport areas enables us to view football, baseball, and other sports at night. Television would not be possible without electricity.

The cases are endless, and almost everyone today realizes the value of better lighting. This field also provides some of the most fascinating and enjoyable work in any branch of the electrical engineering profession.

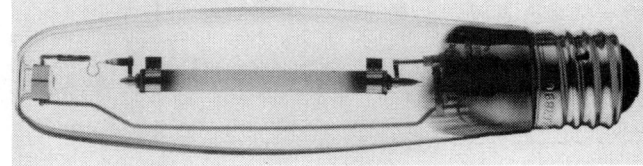

Courtesy General Electric Co.
Fig. 1-4. A modern Lucalox lamp, with a life of 6000 to 8000 hours.

UNIT 2

The Eye and Vision

Since the effect of light upon the eye gives us the sensation of sight, any study of lighting must begin with a consideration of the eye and the seeing process. An understanding of the eye mechanism will help the reader understand the primary function of illumination—to provide light for the performance of visual tasks with a maximum of comfort and a minimum of strain and fatigue.

THE SEEING MECHANISM

The human eye is a fine precision instrument which is often compared to a camera (Fig. 2-1). Both the eye and the camera have a *covering or housing*. Each has a *lens* which focuses an inverted image on a light-sensitive surface—the *retina* in the eye and the *film* in a camera. The camera *shutter* corresponds to the *eyelid*. In front of the lens in the camera is a *diaphragm*, which may be opened or closed to regulate the amount of light entering the camera. The *iris* of the eye performs the same function.

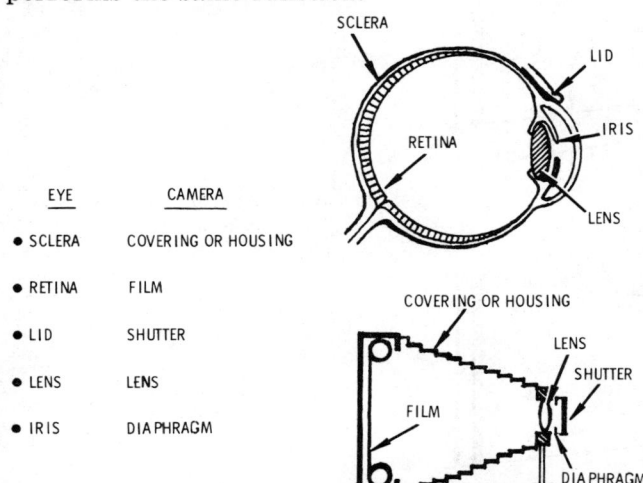

Fig. 2-1. Comparison of the eye to the camera.

There are, however, some important differences between the eye and the camera. The most important being the fact that the eye is a living organ. Taking pictures in poor-quality light will do no harm to the camera. But using the eyes under light of poor quality will result in unnecessary fatigue and may lead to headaches and inflammation of the eyes. Consistent misuse of the eyes can cause permanent damage to them and may also contribute to the development of disorders in other parts of the body. The reader should now begin to realize the importance of proper lighting design for visual tasks.

HOW WE SEE

When a beam of light passes through the transparent protective outer layers of the eye, it is bent or refracted. The amount of light coming through the eye is controlled automatically by the contraction or expansion of the iris. The light continues on through the lens, which focuses the rays through the remaining space of the eyeball and on to the retina. From this point on, the process is electrochemical. Pulsations are set up and are carried to the optic nerve, which, in turn, transmits them to the brain where they are interpreted as light, or where they cause the sensation of sight. Thus, the brain and the eye working together trnsform radiant energy (light) into the sensation of sight.

OBJECTIVE FACTORS IN THE PROCESS OF SEEING

Investigation has shown that the quality of sight depends upon four primary conditions associated with the visual object in question:

1. Size of object.

PRINCIPLES OF ILLUMINATION

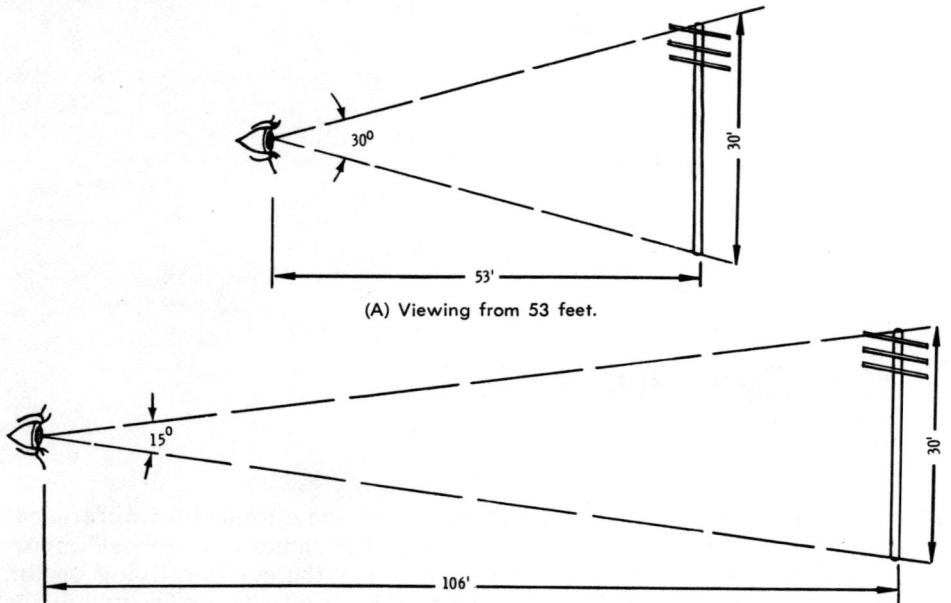

Fig. 2-2. Viewing a 30-foot telephone pole.

(A) Viewing from 53 feet.

(B) Viewing from 106 feet.

2. Luminance (Brightness) of object.
3. The luminance contrast between the object and its immediate background.
4. The time available for seeing the object.

Size

The size of the object is the most generally recognized and accepted factor in seeing. Everyone is familiar with the conventional eye test chart which is used by schools, optometrists, and others for testing visual defects. The larger the object in terms of *visual angle,* the better it can be seen. The following will illustrate this principle.

Fig. 2-2A illustrates a 30-foot telephone pole approximately 53 feet away from the person viewing it. The angle of sight formed from the eye to the object is 30°. If the viewer then backs away from the telephone pole another 53 feet, (Fig. 2-2B), the object has not become smaller; it is still a 30-foot telephone pole. However, since the angle from the eye to the object has become smaller, the object will appear smaller than it did in Fig. 2-2A and cannot be seen as clearly.

To further illustrate, if we take two objects, one 12 inches high and the other 1 inch high (Fig. 2-3A) and place both objects at the same distance

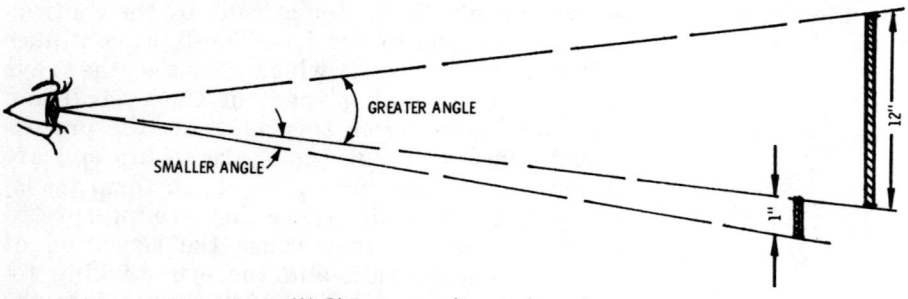

(A) Objects same distance from viewer.

Fig. 2-3. Viewing two objects of different size.

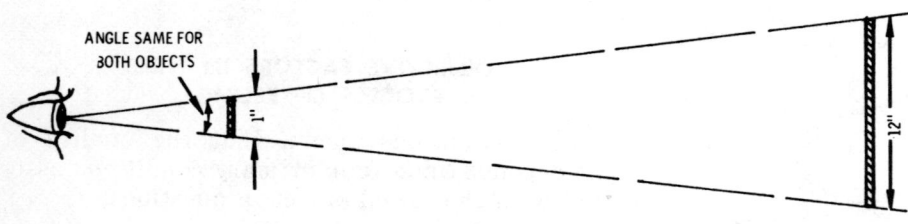

(B) Objects different distances from viewer.

14

from the viewer, the larger of the two objects can be seen more clearly since the angle from the eye to the object will be greater for the larger object.

However, if we move the smaller object closer to the viewer, so that the angle becomes the same as the larger object (Fig. 2-3B), both objects can then be viewed with the same clarity.

Luminance (Brightness)

The brightness of an object depends upon the amount of light striking it, and the proportion of that light reflected in the direction of the eye. A light surface will have a much higher brightness than a dark surface. However, by adding enough light to the dark surface it is possible to make it as bright as the light object. Since the brighter object will be seen first, it will require a lesser amount of light than the darker object for good visibility. Thus, it would take a greater amount of general illumination to adequately light a classroom with walls painted a dark color than it would for one with walls painted a light pastel color.

Fig. 2-4. Contrast between the object and its immediate background.

Contrast

The contrast in brightness or color between the visual object and its immediate background is as important for sight as the general brightness is. The difference in the visual effort required to read the two halves of the circle in Fig. 2-4 demonstrates this fact.

Again, higher levels of illumination will partly compensate for brightness contrast where such conditions cannot be avoided.

Time

Seeing is not an instantaneous process; it requires *time*. The eye can see very well under low levels of brightness if sufficient time is allowed. If the object must be seen quickly, then more light is required. In fact, high lighting levels actually make moving objects appear to move slower, and this greatly increases their visibility. It then stands to reason that it would take a greater amount of illumination to properly light a baseball field than it would for a football field, since the baseball usually will be traveling at a higher rate of speed than the football will.

Size, brightness, contrast, and time are mutually interrelated and interdependent. A deficiency in one can usually be corrected, within limits, by an adjustment in one or more of the others. Of these four conditions, brightness and contrast are usually under the direct control of the lighting designer. With proper control of brightness and contrast, unfavorable conditions, such as size of the object and time given for seeing this object, can be overcome.

SUMMARY

- The purpose of lighting is to make vision possible.
- The mechanism of the human eye is similar to that of a camera.
- Proper illumination is necessary to protect the eyes.
- Good lighting can do much to relieve the eyestrain involved in the performance of difficult visual tasks.
- Research reveals that the advantages of high illumination are even more advantageous to older eyes than to young, normal eyes.
- Size, brightness, contrast, and time are the four basic conditions considered when evaluating the quality of sight.

Only the simple and basic principles of the seeing process have been covered in this unit. However, this should be sufficient background for the material which will follow in future units.

UNIT 3

Characteristics and Measurement of Light

Unit 2 described how the human eye makes use of light in producing the sensation of sight. Thus, for the purpose of our study, light may be defined as a radiant energy evaluated in terms of its capacity for producing the sensation of sight.

All light travels in a straight line unless it is modified or redirected by means of a reflecting, refracting, or diffusing condition.

LIGHT COLORS

The different colors of light are due to the different wave frequencies, which are considered to be of an electromagnetic nature, and are known to be of extremely high frequency and much shorter in length than the shortest television waves.

Ordinary sunlight, while it appears white, is actually made up of a number of colors. In 1666, Sir Isaac Newton passed a beam of light through a prism and discovered that it contained all colors of the rainbow. The three basic colors are red, blue, and green, but by continuously blending together, they also produce violet, yellow, and orange (Fig. 3-1).

Artificial white or daylight is generally the most desirable form of light for illuminating purposes, but it must contain a certain number of the colors which compose sunlight. It is the reflection to our eyes of these various colors from the object they strike that enables us to see objects and to get impressions of their color. Certain surfaces and materials absorb light of one color and frequency and reflect that of another color; this gives us our color distinction in seeing different things.

White and light-colored surfaces reflect more light than dark surfaces do. (See *contrast* in Unit 2.)

The ordinary tungsten-filament electric lamp is a good example of nearly white artificial light that is excellent for most applications. The molecules of the tungsten wire are caused to vibrate rapidly and produce heat when an electric current is applied. When enough current is passed through the wire, it becomes incandescent (white light).

UNITS OF LIGHT MEASUREMENT

Before we can undertake to perform actual lighting layouts or select equipment for certain applications, it is necessary to learn more about actual quantities of light, units of light, etc. An understanding of these interesting units and principles will help us to better understand the nature of light.

When we go to purchase an incandescent lamp, we normally refer to the rating of the lamp in terms of *watts* (a 60-watt lamp, 150-watt lamp, etc.). While the rating in watts will usually give us a general idea of the lamp size, it does not tell how much light a certain lamp can be expected to produce.

For example, one might expect a 100-watt incandescent lamp to produce more light than a 40-

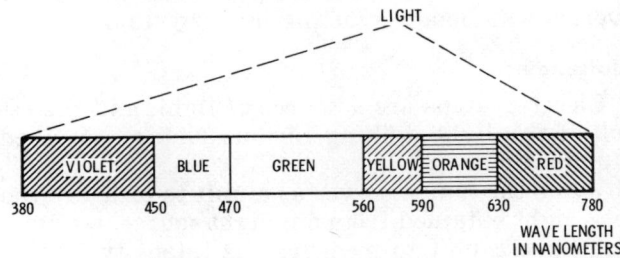

Fig. 3-1. The visible light spectrum.

PRINCIPLES OF ILLUMINATION

watt fluorescent lamp. However, the average inside-frosted 100-watt incandescent lamp emits light at the rate of about 1490 lumens, while the flow of light from a 40-watt fluorescent lamp is about 3200 lumens.

The preceding example implies that the total amount of light actually given off by a light source is measured in terms of the unit *lumen*.

Lumen

A lumen may be defined as the quantity of light which will strike a surface of one square foot, all points of which are one foot distance from a light source of one candlepower (one standard candle for our purposes).

A lumen of light may be visualized by placing a standard candle inside a hollow sphere which has a radius of one foot or a diameter of two feet, and the inside of which is completely black to prevent any reflection of light. If a one foot square is cut out of the sphere, as shown in Fig. 3-2, the amount of light that will escape through this hole will be one lumen.

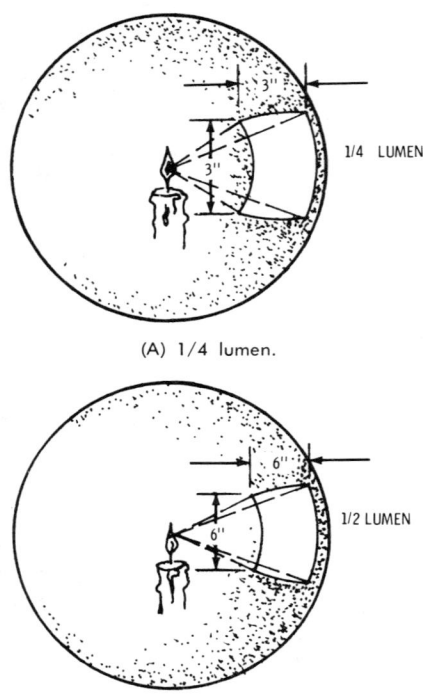

(A) 1/4 lumen.

(B) 1/2 lumen.

Fig. 3-3. Using a sphere to illustrate a portion of a lumen.

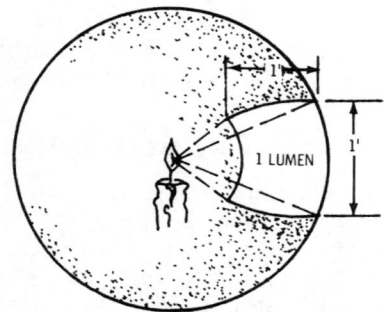

Fig. 3-2. Using a sphere to illustrate one lumen.

If the area of the hole was ¼ square foot, then the light emitted from the hole would be ¼ lumen; if the hole was ½ square foot, the escaping light would be ½ lumen and so on (Fig. 3-3). A sphere with a 1-foot radius has a total area of 12.57 square feet, so if the entire sphere was removed from the candle, the total lumens emitted by the standard candle would be 12.57. From this we find that we can determine the approximate amount of lumens given off by any lamp by multiplying the average candlepower of the lamp by 12.57.

Footcandles

Electric lamps are a source of light, and the result of this light striking various surfaces is called illumination.

While the lumen serves as a unit to measure the total light obtained from any light source, we must also have a unit to measure the intensity of this light on a given surface. (Desk tops and work benches would be good examples of such surfaces.) The unit used for this purpose is called the *footcandle*.

A footcandle is a unit of measurement which represents the intensity of illumination that will be produced on a surface that is one foot distance from a source of one candlepower, and at right angles to the light rays from the source.

The footcandle, then, is the unit used in illumination problems to determine the proper level of illumination on any working plane or surface.

Referring to Fig. 3-4, we find that the surface A, B, C, D is illuminated at every point with an intensity of one footcandle. We previously learned that the total amount of light striking this same surface is also one lumen. Thus, when one lumen of light is evenly distributed over a surface of one square foot, that same area is illuminated to an intensity of one footcandle.

The preceding paragraph shows that if we know the area of a surface that is to be lighted and the

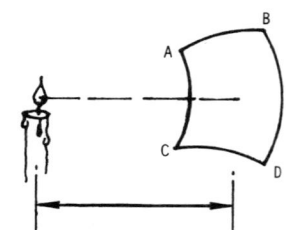

Fig. 3-4. Illustration of one footcandle.

CHARACTERISTICS AND MEASUREMENT OF LIGHT

intensity in footcandles of the desired illumination level, we can then calculate the number of lumens that will be required to light the area. For example, if we desire to illuminate a surface of 100 square feet to an average intensity of 10 footcandles by a light source one foot distance from the surface, we multiply the desired footcandles by the area of the surface to be illuminated—10 × 100. Therefore, the source of light must produce 1000 lumens in order to light the surface to an intensity of 10 footcandles. In actual practice, more lumens would be required to produce 10 footcandles of illumination on this surface, since the efficiency of any light source is rarely 100%. More light will also be required as the distance between the light source and surface is increased.

Inverse Square Law for Light

Footcandle units are used to indicate the illumination level at a specific point, or the average illumination on a surface or working plane. The inverse square law is the basis of calculation in the point-by-point method of lighting design as described in Unit 10.

The inverse square law states that the illumination on a surface varies directly with the candlepower of the source of light, and inversely with the square of the distance from the source. This law informs us that a small change in distance from a light source will make a great change in the illumination level on a surface. Fig. 3-5 illustrates the reasons for this.

In Fig. 3-5 we have a light source of one candlepower, and since the surface at "A" is one foot from the candle, its illumination intensity will be one footcandle. If we move the surface or plane to "B," which is two feet from the source, the same number of light rays will have to spread over four times the area, as that area increases in all directions. Then the illumination intensity at twice the distance is only ¼ the amount it was before, as the distance of two squared is four, and this is the number of times the illumination is reduced.

If we continue to move the surface to "C," which is three feet away from the light source, the rays are now spread over 9 times the original area ($3^2 = 9$), and the intensity of illumination on the surface will now be only 1/9 of its former value.

The farther away any surface is from a source of light, the less light it receives from that source, since the same light rays must be distributed over a greater area.

Various models of convenient portable footcandle meters are available to measure the footcandle level on surfaces or planes. A discussion of these instruments and their uses is found later in this unit under field measurements.

LIGHT REFLECTION

We all know that light can be reflected from certain light-colored or highly polished surfaces. This fact is taken into consideration when performing lighting calculations.

Some surfaces and materials are much better reflectors than others. Usually, the lighter colors reflect more light and absorb less light than darker colors.

The percentages of light that will be reflected from some of the more common materials are as follows:

White plaster	90% to 92%
Mirrored glass	80% to 90%
White paint	75% to 90%
Metallized plastic	75% to 85%
Polished aluminum	75% to 85%
Stainless steel	55% to 65%
Limestone	35% to 65%
Marble (white)	45%
Concrete	40%
Dark red glazed bricks	30%

The better classes of reflectors are used in directing light. The colors of walls, ceilings, and floors and their reflecting ability are also considered in interior lighting design. Surface reflectance will be more fully covered in later units.

FIELD MEASUREMENTS

There are a number of large and elaborate devices used in laboratories for making exact tests and measurements on light and lighting fixtures. But for practical use in the field, a portable light or footcandle meter, as illustrated in Fig. 3-6, is

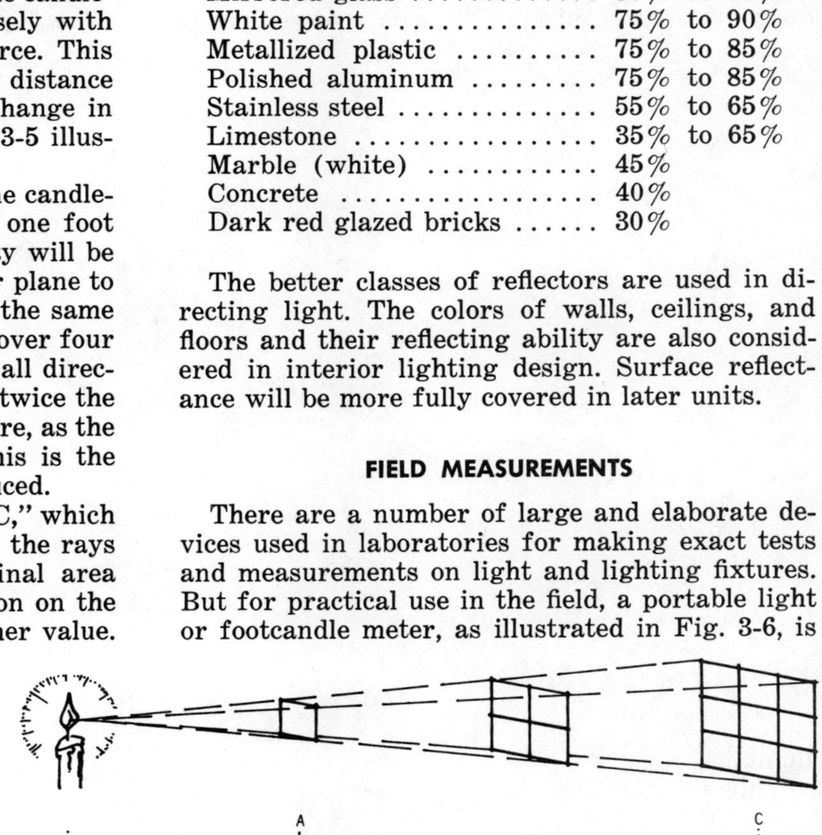

Fig. 3-5. The inverse square law illustrated.

quite satisfactory. This type of footcandle meter can be purchased for around $10.

The footcandle meter illustrated in Fig. 3-6 consists of light barrier layer cells and a meter, enclosed in a suitable covering. This meter will read intensities from 1 to 500 footcandles.

Fig. 3-6. A typical footcandle meter.

To use a footcandle meter of the type previously described, first remove the cover. Hold the meter in a position so the cell is facing toward the light source and at the level of the work plane where the illumination is required. The shadow of your body should not be allowed to fall on the cell during tests. A number of such tests at various points in a room will give the average illumination level in footcandles.

SUMMARY

- While ordinary sunlight appears white, it is actually made up of a number of colors: violet, blue, green, yellow, orange, and red.
- Artificial white or daylight is generally the most desirable form of light for illuminating purposes.
- White and light-colored surfaces reflect more light than dark surfaces do.
- The total amount of light actually given off by a light source is measured in terms of the unit *lumen*.
- The level of illumination on any surface or work plane is measured in terms of the unit *footcandle*.
- The instrument used to measure the illumination level is called the *footcandle meter*.
- The farther any surface is from a source, the less light it receives from that source.

SECTION II

LIGHT SOURCES

UNIT 4

Lamp Classifications

Now that we know something of the nature of light and the fundamentals of good illumination, we can discuss the three common sources of electric light:

1. Incandescent lamps.
2. Gaseous discharge lamps.
3. Electroluminescent lamps.

Despite continuous improvement, none of these light sources have a high overall efficiency. The very best light source converts only approximately ¼ of its input energy into visible light. The remaining input energy is converted to heat or invisible light. Fig. 4-1 illustrates the energy distribution of a typical cool white fluorescent lamp.

Incandescent Lamps

Incandescent lamps are made in thousands of different types and colors, from a fraction of a watt to over 10,000 watts each, and for practically any conceivable lighting application.

Extremely small lamps are made for instrument panels, flashlights, etc., while large incandescent lamps, over twenty inches in diameter, are used for spotlights and street lighting.

Regardless of the type or size, all incandescent filament lamps consist of a sealed glass envelope containing a filament (Fig. 4-2). The incandescent filament lamp produces light by means of a filament heated to incandescence (white light) by its resistance to a flow of electric current. Most of these elements are capable of producing 11 to 22 lumens per watt, and some produce as high as 33 lumens per watt.

The filaments of incandescent lamps were originally made of carbon. Now, tungsten is used for virtually all lamp filaments because of its higher melting point, better spectral characteristics, and strength—both hot and cold.

Tungsten-Halogen Lamps

The quartz-iodine tungsten-filament lamp is basically an incandescent lamp, since light is produced from the incandescence of its coiled tungsten filament. However, the lamp envelope, made of quartz, is filled with an iodine vapor which prevents the evaporation of the tungsten filament. This evaporation is what normally occurs in conventional incandescent lamps. When the bulb be-

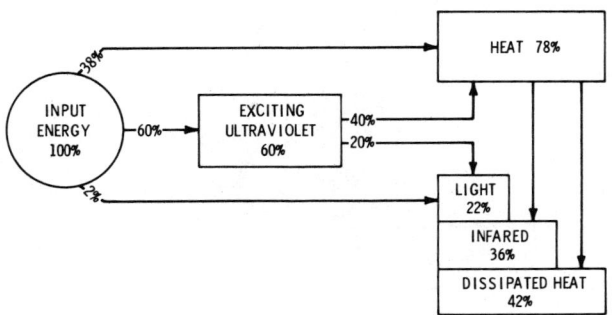

Fig. 4-1. Energy distribution of a typical cool-white fluorescent lamp.

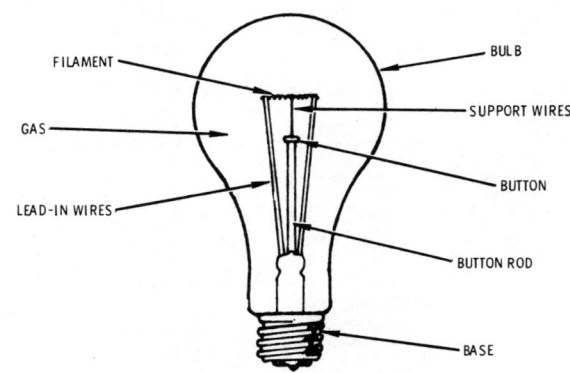

Fig. 4-2. An incandescent lamp.

PRINCIPLES OF ILLUMINATION

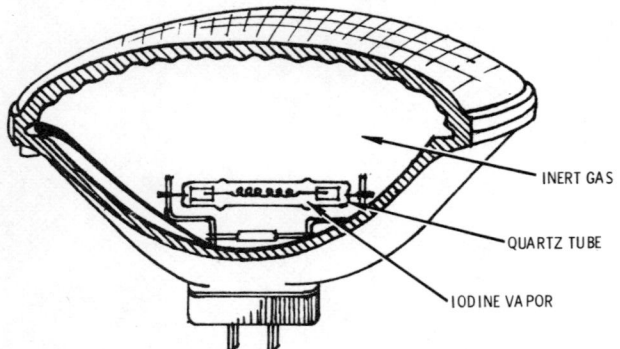

Fig. 4-3. The basic components of a quartz-iodine lamp.

gins to blacken, light output deteriorates, and eventually the filament burns out. While the quartz-iodine lamp has approximately the same efficiency as an equivalent conventional incandescent lamp, it has the advantages of double the normal life, low lumen depreciation, and a smaller bulb for a given wattage. Fig. 4-3 illustrates the basic components of a quartz-iodine lamp.

Electric Discharge Lamps

The electric discharge lamp category includes the well-known fluorescent, neon, and mercury-vapor lamps as well as the newer metal-halide and sodium lamps. In this group of lamps, light is produced by the passage of an electric current through a tungsten filament. The application of an electrical potential ionizes the gas and permits current to flow between two electrodes located at opposite ends of the lamp. This arc discharge accelerates to tremendous speeds, and when the current collides with the atoms of the gas it temporarily alters the atomic structure. Light results from the energy given off by these altered atoms as they return to their normal state.

The incandescent lamp has certain characteristics which make it inherently inefficient as a source of light; maximum possible values for this type of lamp have probably already been approached. However, the electric discharge lamp produces light by an entirely different process and is capable of achieving a much higher efficiency of light output.

Fluorescent Lamps

Of all the electric discharge light sources, the fluorescent lamp is the best known and most widely used (Fig. 4-4). Since fluorescent lighting was introduced to the general public, during the 1933 Chicago Centennial Exposition, it has almost completely taken the place of the incandescent lamp in all branches of construction, except for specialty lighting and possibly for residential use.

One of the reasons for the popularity of fluorescent lighting is its high efficiency as compared to incandescent lamps. The average 40-watt incandescent lamp delivers approximately 470 initial lumens, while a 40-watt fluorescent lamp delivers about 3150 lumens. This power efficiency not only saves on the cost of power consumed, but it lessens the heat and reduces air conditioning loads (another saving). Fluorescent lighting allows more comfort for those working under bright lights during warm weather. Other advantages will be discussed in Unit 7.

Fluorescent lamps are made with long glass tubes sealed at both ends and containing an inert gas, generally argon, and low pressure mercury vapor. Built into each end is a cathode which supplies the electrons to start and maintain the gaseous discharge.

The inside of the lamp tubing is coated with a thin layer of materials called phosphors. A phosphor is a substance which becomes luminous or which glows with visible light when struck by streams of electrons, which are caused to pass through the space between filaments inside the lamp. When the phosphors are thus made luminous, the action is called fluorescent, which gives the lamp its name.

Mercury Lamps

This type of lamp has been used mainly for outdoor lighting and for industrial applications due to its poor color and high wattage. However, re-

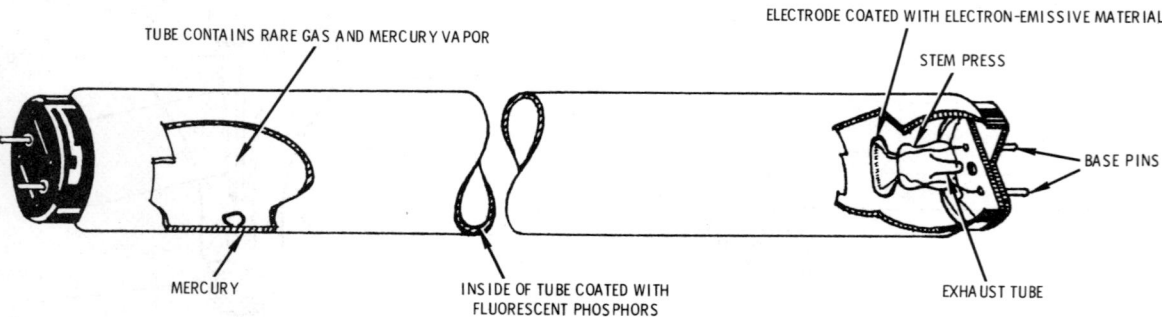

Fig. 4-4. A fluorescent lamp.

cent improvements in color, availability in lower wattages, and a higher output efficiency have made these lamps attractive for general indoor application. See Fig. 4-5 for a typical mercury-vapor lamp.

The traditional mercury-vapor lamp produces light with a predominance of yellow and green rays and a small percentage of violet and blue.

Since this blue-green light distorts almost all colors, color correction has been added by coating the outer bulb with phosphors. The phosphors are activated by the ultraviolet light and reradiate generally in the red bond, which is entirely absent in the basic lamp color. The light has now been corrected to make it acceptable for indoor use.

Mercury-vapor lamps operate by passing an arc through a high-pressure mercury vapor that is contained in an arc tube made of quartz or glass. This action produces visible light. As with all arc discharge lamps, ballasts are required to start the lamp, and thereafter to control the arc.

Metal-Halide Lamp

The metal-halide lamp is basically a mercury lamp which has been altered by adding iodine compounds to the mercury and argon gas in the arc tube. These iodine compounds are of metals such as indium, sodium, thallium, or dysprosium. The addition of these salts causes the emission of light which is of a better color than the basic mercury colors, although life and lumen maintenance may be decreased in the process.

Metal-halide lamps furnish approximately 75 to 90 lumens per watt of white light, which is much warmer than the mercury light and is suitable for all indoor applications.

High-Pressure Sodium-Vapor Lamps

The high-pressure sodium-vapor lamps use an arc tube of ceramic material such as polycrystalline alumina. The lamp operates in a similar manner to the other discharge lamps, producing a warm yellow-orange tone at the rate of over 100 lumens per watt. This type of lamp is mostly used for outdoor applications.

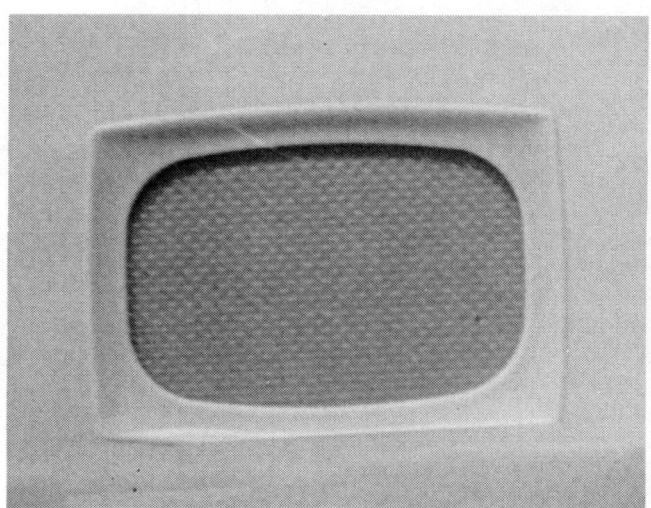

Fig. 4-6. An electroluminescent night light.

Unit 6 will fully explain the operating principles of all electric discharge lamps, along with the recommended use of each.

Electroluminescent Lamps

The fluorescent lamp produces light by exciting phosphors with ultraviolet light. With certain special phosphors, it is possible to produce light directly from electric energy without the use of the intermediate step. This process is known as electroluminescence.

Though still in early stages of development, this process has been successfully used in night lights (Fig. 4-6), road sign illumination, control panel lighting, and other applications requiring low brightness and maintenance-free illumination.

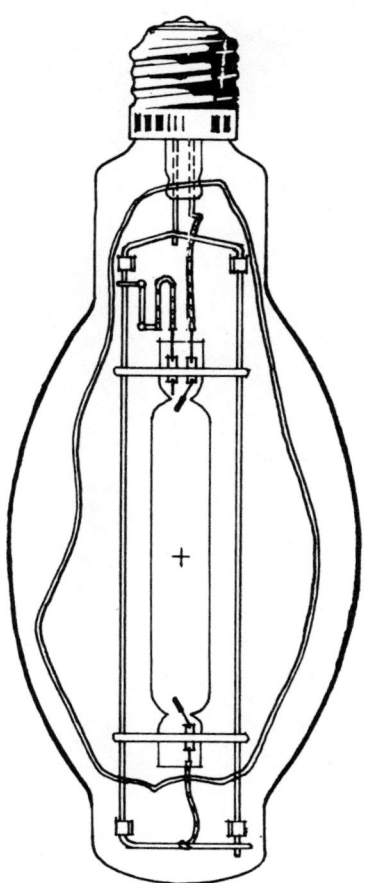

Fig. 4-5. A mercury-vapor lamp.

SUMMARY

- Common sources of electric lamps generally fall into three categories:
    1. Incandescent lamps.
    2. Electric discharge lamps.
    3. Electroluminescent lamps.

- Incandescent lamps produce light by passing an electric current through a high resistance tungsten filament until it heats to incandescent (white light).

- Electric discharge lamps produce light by passing an electric current through a vapor or gas, rather than through a tungsten filament.

- Electroluminescent lamps convert electrical energy directly into light without any intermediate steps.

UNIT 5

Incandescent Lamps

Electric incandescent filament lamps consist of three major parts (Fig. 5-1): bulb, base, and filament.

BULB

The sealed glass envelope enclosing the filament is often called the bulb and is used to obtain a vacuum or an atmosphere of inert gas. Without such an atmosphere, the filament would rapidly disintegrate due to oxidation.

The filaments of all early incandescent lamps operated in a vacuum—all air and gas, insofar as practical, was exhausted from the space within the bulb and surrounding the filament. In a vacuum lamp, the heat losses by convection and conduction are reduced, but the filament begins vaporizing at a lower temperature and therefore evaporates more rapidly than it would if pressure was applied.

The purpose of gas inside the bulb is to create pressure on the filament to retard evaporation, and this type of lamp is considered more efficient than a vacuum lamp of the same size. Since the filament can operate at a higher temperature in a gas-filled lamp, it also produces a whiter light than that produced by a vacuum lamp of the same size.

Bulb Size and Shape

Bulbs are made in a variety of shapes and sizes for use in a large variety of applications. Typical bulb shapes are shown in Fig. 5-2.

The sizes and shapes of lamp bulbs are designated by a letter or letters followed by a number. The letter indicates the shape of the bulb: S, straight side; F, flame; G, round or globular; T, tubular; A, arbitrary designation applied to the

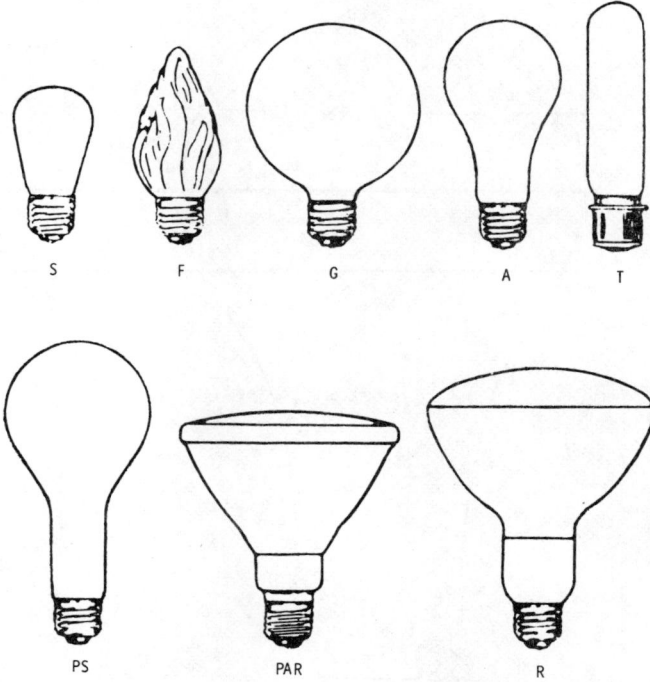

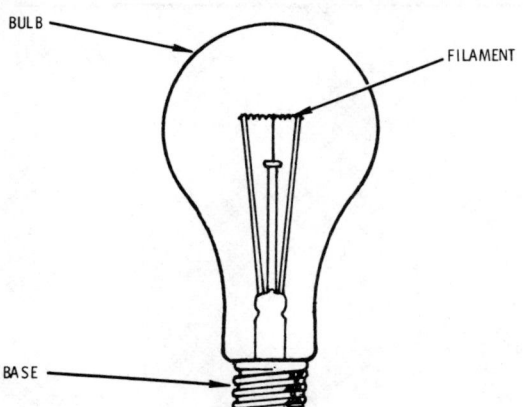

Fig. 5-1. The major parts of an incandescent filament lamp.

Fig. 5-2. Some typical bulb shapes.

27

PRINCIPLES OF ILLUMINATION

bulbs commonly used for general lighting service lamps of 200 watts or less.

The number in a bulb designation indicates the maximum diameter of the bulb in eighths of an inch. For example, an A-21 bulb is 21 eighths of an inch or 2⅝ inches in diameter at its maximum dimensions. To determine the size of a bulb with no markings, the full size scale in Fig. 5-3 (calibrated in ⅛ inches) may be used. Place one edge of the bulb to be identified on the "index" mark (0), and measure the approximate diameter by reading the scale at the opposite edge of the bulb. The broken line outlining an A-19 lamp illustrates this procedure.

Bulb Finish and Color

To diffuse the light from the filament, many lamps have inside-frosted bulbs, produced by a light acid etching applied to the inner surface of the bulb. Some types of lamps are available with an inside white silica coating which provides still greater diffusion. The inside-frosted bulb absorbs no measurable amount of light, whereas the silica coating absorbs about 2%. With both treatments the outer surface of the bulb is left smooth and easily cleaned. Diffusing bulbs are preferred for most general lighting purposes, but where accurate control of light is involved, as in optical systems, clear bulb lamps are necessary.

Other finishes applied to some general lighting service lamps are white bowl and silvered bowl. A white-bowl lamp has a translucent white coating on the inner surface of the bulb bowl, which serves to reduce both direct and reflected glare from open fixtures. A silvered-bowl lamp has an opaque silver coating applied to the bowl. The inner surface of this coating is a highly specular reflector which is not affected by dust or deterioration, and therefore remains efficient throughout the life of the lamp. Silvered-bowl lamps are commonly used in certain types of equipment for totally indirect lighting, and also occasionally in direct fixtures such as standard dome reflectors.

Colored light in filament lamps is produced subtractively, by means of a bulb that absorbs light of colors other than that desired. Most colored bulbs are made by applying a pigmented coating to either the inner or the outer surface of a clear bulb, or by fusing an enamel into the outer surface (ceramic coating). The colors in most common use are red, blue, green, yellow, orange, ivory, flame-tint, and white.

Lamps with a slightly pink-colored inside silica coating (Beauty-Tone lamps) are available in the three-way design. These lamps are primarily used in residential lighting equipment. They are used where delicately tinted light is desired for a decorative effect. Ceramic coatings and inside coatings are satisfactory for either outdoor or indoor use, but most outside coatings are not permanent and are recommended for use only where they are protected from the weather.

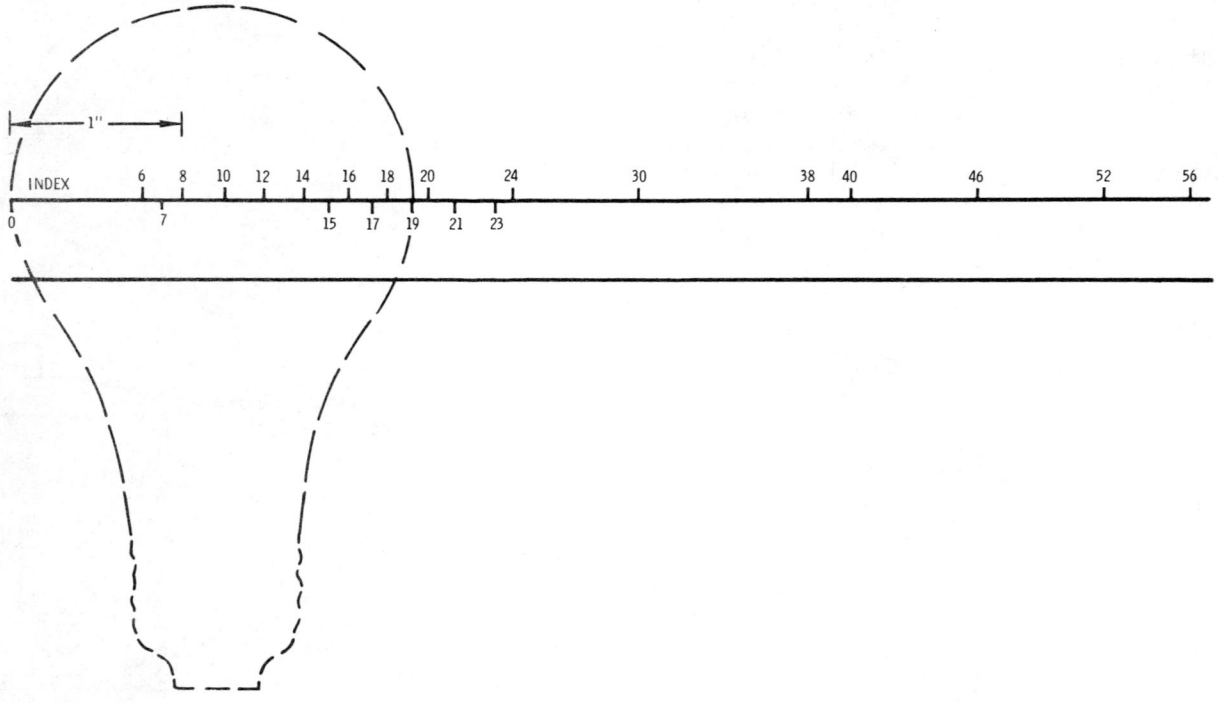

Fig. 5-3. A bulb size scale.

INCANDESCENT LAMPS

Another type of colored lamp has a bulb of natural-colored glass, made by adding chemicals to the ingredients of the glass. Natural-colored bulbs are made in daylight blue, blue, amber, green, and ruby. They produce light of purer colors than coated bulbs and are often used for theatrical and photographic lighting purposes. Where decorative or display lighting is involved, coated lamps are preferred to natural-colored lamps because of their lower cost.

The most widely used of the natural-colored lamps is the daylight blue. The characteristics of the daylight blue bulb are such as to reduce the preponderance of red and yellow light common to incandescent lamps, with the result that the light produced more nearly approaches daylight in color. Since this is accomplished at the expense of increased lamp cost and of some 35% absorption in light, daylight blue lamps should be used only where the lighting requirements make it necessary.

BASE

The base provides a means of connecting the lamp bulb to the socket. For general lighting purposes, screw-type bases are most commonly used. Most general lighting service lamps (300 watts and below) have medium screw bases. The higher wattages (300 watts and above) use the mogul screw base. Some of the lower-wattage lamps, particularly the sign, indicator, and decorative types, have candelabra or intermediate screw bases. See Fig. 5-4 for incandescent base illustrations.

A light source (lamp filament) cannot be accurately aligned with respect to an optical system by means of a screw base. Filament orientation is provided by a number of other types of bases. The most common bases are the prefocus, bipost, bayonet, and special pin-type bases for projection lamps. A bipost base, usually used on high-wattage lamps, consists of two metal pins or posts imbedded in a glass "cup" forming the end of the lamp bulb. Most screw and prefocus bases are attached to the bulb by means of a basing cement especially designed for the purpose. Other bases used on certain lamps include prong types, screw terminals, contact lugs, flexible leads, recessed single contact, and a number of other types for specific applications.

FILAMENT

The filament is the light-producing element of the lamp, and the primary considerations in its design are its electrical characteristics. The wattage of a filament lamp is equal to the voltage delivered at the socket times the amperes flowing through the filament. By Ohm's law the current is determined by the voltage and by the resistance, which in turn depends on the length and the diameter of the filament wire. The higher the wattage of a lamp of a given voltage, the higher the current, and therefore the greater the diameter of the filament wire required to carry it. The higher the voltage of a lamp of a given wattage, the lower the current and the smaller the diameter of the filament wire.

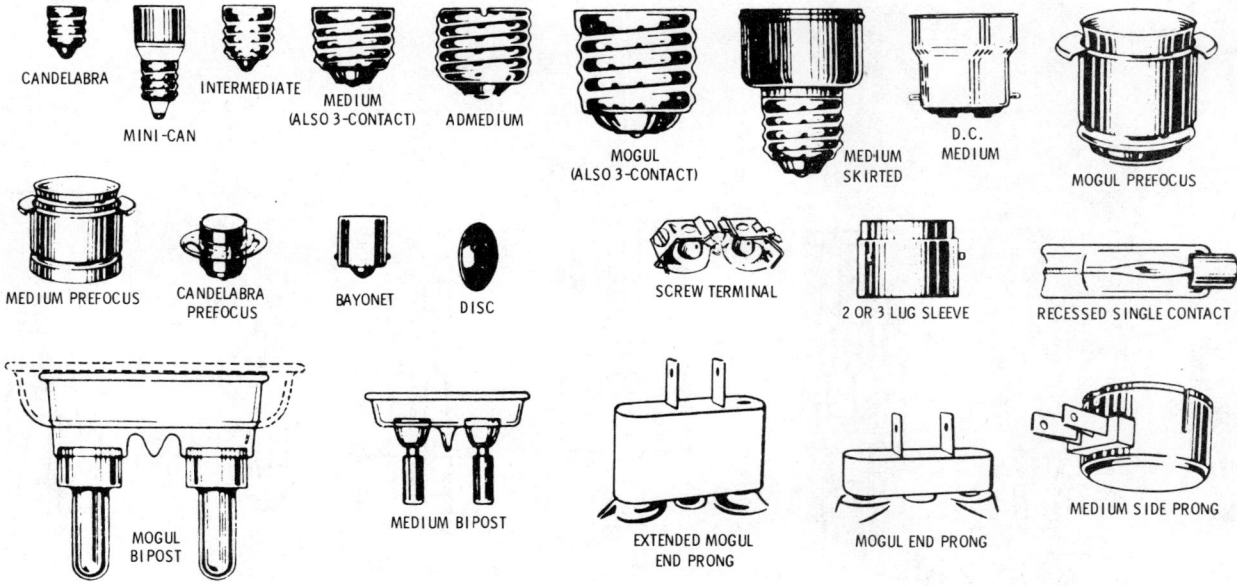

Courtesy General Electric Co.

Fig. 5-4. Incandescent base illustrations.

PRINCIPLES OF ILLUMINATION

The higher the operating temperature of the filament, the greater the share of the emitted energy that lies in the visible region of the radiation spectrum. Since most filament lamps radiate as light only about 10 to 12% of the input energy, it is important to design a lamp for as high a filament temperature as is consistent with satisfactory lamp life. Carbon, which has a higher melting point than tungsten and was one of the early filament materials, has been almost completely replaced by tungsten because carbon at high temperatures evaporates too rapidly, whereas tungsten combines the properties of high melting point and slow evaporation.

Since the larger the diameter of the filament wire the higher the temperature at which it can be operated without danger of excessive evaporation, high-wattage lamps are more efficient than low-wattage lamps of the same voltage and life rating. A 150-watt 120-volt general lighting service lamp, for example, produces 34% more light than three 50-watt 120-volt lamps consuming the same wattage. It also follows that low-voltage lamps, because their filament wire diameter is greater, are more efficient than higher-voltage lamps of the same wattage.

The filament forms in common use today (Fig. 5-5) are designated by a letter or letters indicating whether the wire is straight or coiled, a number specifying the general form of the filament, and sometimes another letter indicating arrangement on the supports. "S" as the first letter of a filament designation means a straight (uncoiled) filament wire, "C" a coiled wire, "CC" a coiled coil, and "R" a flat or ribbon-shaped wire. The numbers and other letters assigned to the various filament forms are purely arbitrary.

Early lamps were made with straight filaments operating in a vacuum. When inert gases were introduced into the bulb, it was found that coiling the wire decreased the effective surface exposed to the circulating gas, and therefore reduced the heat lost by conduction and convection. The coils also tend to heat each other, and the coiled filament is mechanically stronger. Today nearly all types of lamps, both vacuum and gas-filled, have coiled

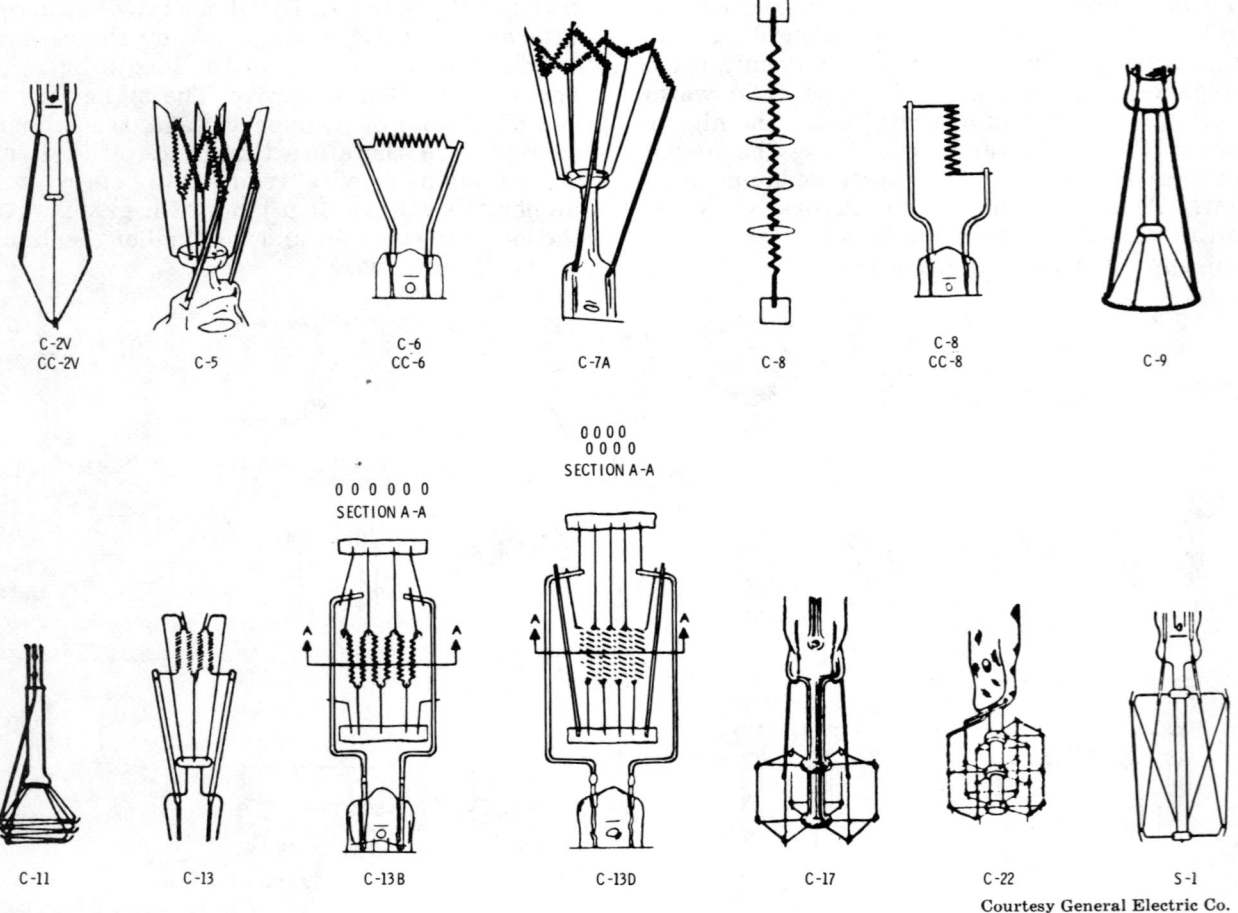

Fig. 5-5. Typical filament forms.

INCANDESCENT LAMPS

filaments. The single-coil filament is formed by winding the tungsten wire on a mandrel of steel or molybdenum in a continuous process. The coil with the mandrel still in place is cut into the desired lengths and immersed in an acid bath, which dissolves the mandrel but does not attack the tungsten.

Coiled-coil, or double-coiled, filaments which provide increased efficiency and reduced light-source size are at present used in various general lighting service standard-voltage lamps in the 50- to 1000-watt range, also in certain types of projection lamps. The process of making coiled-coil filaments is the same as that for making single-coil filaments. With the mandrel intact, the wire is wound onto another mandrel which is later "retracted," or removed mechanically. The first mandrel is then removed from the coiled coil by dissolving.

In the general lighting service type of lamp, the arrangement of the filament coil and its supports is dictated by the limiting size of the bulb neck through which it must be inserted, and by other manufacturing considerations. Mounting a filament vertically rather than horizontally (C-8, CC-8, or 2CC-8 construction), as has recently been done in general lighting service lamps, results in a higher light output because gas convection currents raise the filament temperature and because less light is absorbed by the lamp base. Further, the bulb blackening which develops as the lamp ages is localized within a smaller area, and lumen maintenance throughout the lamp life is higher. Lamps for special purposes often require certain filament forms. For projection, searchlight, spotlight, floodlight, and similar services where accurate control of light demands a small source, the filaments are concentrated into as small a space as possible. In contrast, for showcase service where a long light source is needed, the filament may be extended along almost the full length of the bulb.

FILLING GAS

Incandescent lamps were first made with evacuated bulbs, the purpose being merely to keep the filament from burning up by excluding oxygen. Later it was discovered that the pressure exerted on the filament by an inert gas introduced into the bulb retarded the evaporation of tungsten, thus making it possible to design lamps for higher filament temperatures. Vacuum lamps are now designated as "type B" lamps, gas-filled lamps as "type C."

The gas removes some heat from the filament, as a result of conduction and convection losses not present in the vacuum lamp. The larger the surface of the wire in proportion to its volume or mass, the greater this cooling effect becomes, until eventually it nullifies the gain achieved by using the filling gas. Filaments with a current rating of less than $\frac{1}{3}$ ampere have a wire diameter so small that the introduction of gas is a disadvantage. For this reason, standard-voltage general lighting service lamps of less than 40 watts are of the vacuum or "type B" construction, while lamps of 40 watts and higher are gas-filled.

Nitrogen and argon are the gases most commonly used in lamp manufacture. Projection lamps use an atmosphere of 100% nitrogen. Most other types have a mixture of nitrogen and argon, the proportions varying with the lamp and the service for which it is designed. High-voltage lamps, for example, are filled with approximately 50% argon and 50% nitrogen, the higher wattage standard-voltage types about 90% argon and 10% nitrogen, and the lower wattage standard-voltage types and all street series lamps about 98% argon and 2% nitrogen. Some nitrogen is necessary to prevent arcing across the lead-in wires, which would occur if pure argon were used. The greater the inherent tendency of a lamp to arc, the higher the percentage of nitrogen in its gas mixture.

Krypton is a relatively rare and expensive gas which has a higher atomic weight than either argon or nitrogen, and therefore causes less energy loss by conduction and convection. It is primarily used in certain miniature lamps such as those on miners' caps, where the limited capacity of the battery power supply makes it essential to obtain the greatest possible efficiency. Hydrogen, because of its low atomic weight, is used in certain very special types of flashing signal lamps where rapid cooling of the filament is important.

TYPES OF LAMPS

General Lighting Service Lamps

The familiar general lighting service lamps, from the 15-watt A-15 to the 1500-watt PS-52, designed for multiple burning on 120-, 125-, or 130-volt circuits, are the most commonly used filament-type lamps. All standard general service lamps are equipped with screw bases. The larger wattages are manufactured in either clear or inside-frosted bulbs. Below 150 watts, inside frosted and inside white silica coated lamps are standard. The wattages most commonly used in the home are available in a straight-sided modified T-bulb shape, with the white silica coating.

High- and Low-Voltage Lamps

Lamps similar to those of the standard-voltage line are available for operation on 230 and 250 volts. The low efficacy of these lamps, as compared to comparable lamps of standard-voltage rating, has already been mentioned. Other disadvantages, resulting from the smaller filament wire diameter of high-voltage lamps, are reduced mechanical strength and larger overall light-source size which makes them less satisfactory for use in floodlight and projection equipment. The only gain achieved by the industrial use of these higher voltages is the reduction in ampere load which results from doubling the voltage, and the consequent saving in wiring cost. Lamps for operation on 30- and 60-volt circuits are also available for use in train lighting and in country home service.

Projector and Reflector Lamps

PAR-bulb (projector) and R-bulb (reflector) lamps combine, in one unit, a light source and a highly efficient sealed-in reflector consisting of vaporized aluminum or silver applied to the inner surface of the bulb. The 100-watt PAR-38 and 150-watt R-40 lamps are available in several colors. "PAR" bulbs are of hard glass. "PAR" lamps up to 150 watts in size, as well as a few special service "R" lamps with heat-resistant-glass bulbs, can be used outdoors without danger of breakage from rain or snow. Larger "PAR" lamps and all other "R" lamps are not recommended for outdoor use unless protected from the elements.

Higher-wattage R-52 and R-57 reflector lamps are designed for general lighting purposes. They are made in both wide and narrow distributions and are best adapted for high-ceilinged industrial areas where the atmosphere contains noncombustible dirt, smoke, or fumes. Where heat-resisting glass is required for protection against thermal shock, the R-60 lamps will perform similarly. These latter types are especially suited for outdoor floodlighting. In addition to flood and spotlight service, PAR-bulb lamps have found wide application in automotive, aviation, and other miscellaneous fields where compact lighting units of precise beam control are necessary.

Showcase and Lumiline Lamps

Low-wattage tubular-bulb lamps are used for showcase lighting and other applications where small bulb diameter is required. Some of these are designed to be used in reflectors, and others are provided with an internal reflecting surface extending over approximately half the bulb area, which concentrates the light to form a beam. The Lumiline lamp is a special type of tubular light source which has a filament extending the length of the lamp. The filament is connected at each end to a disc base which requires a special type of lamp holder. Lumiline lamps are considerably less efficient than conventional general lighting service lamps, but are useful where a linear source is necessary.

Spotlight, Floodlight and Projection Lamps

Characteristic features of all lamps designed for spotlight, floodlight, and projection applications include compact filaments accurately positioned with respect to the base, for purposes of light control; relatively short life, for high efficacy and luminance; comparatively small bulbs; and restricted burning position. Since spotlight lamps must produce narrower, more intense beams than floodlight lamps, they usually have smaller filaments and shorter lives. In projection lamps the light source is still more concentrated and lamp life is further reduced, with accompanying increased efficacy.

The objective in designing projection lamps is to fill the aperture of the projection system with a light source of high luminance and maximum uniformity. This is accomplished by arranging the filament coils in a single or double vertical plane and using a base which accurately locates the filament with respect to the optical system. The biplane (C-13D) filament, with coils arranged in two parallel rows so placed that the coils of one row fill in the spaces between those of the other, has much greater uniformity and higher average luminance than the single-row monoplane (C-13) filament. Many projection lamps have such small bulbs and operate at such high temperatures that they cannot be burned without continuous forced ventilation, and some have designed lives as short as 10 hours. Lamps for use in certain types of projectors have an opaque coating on the top of the bulb to prevent the emission of stray light.

Halogen Lamps

The halogen lamp is a new concept in incandescent lamps. It uses a quartz envelope which is the basis for its many advantages which include: compactness, thermal shock resistance, high efficacy, and almost perfect maintained light throughout the lamp life. Iodine is used in the lamp to create a chemical cycle with the sublimated tungsten to keep the bulb clean. The halogen lamp is used for floodlighting, aviation, photographic, special effects, photocopy, and other applications where its special features are desirable.

Infrared Lamps

Infrared lamps are essentially the same as lamps designed for illumination purposes; the principal difference between them is filament temperature. Since the production of light is not an objective, infrared lamps are designed to operate at a very low temperature, resulting in a low light output (about 7 or 8 lumens per watt) and a consequent reduction in glare. If only the advantage of low filament evaporation is desired, the life of infrared lamps is many thousands of hours; but because of the possibility of failure from shock, vibration, and other causes, the rated life is given merely as "in excess of 5000 hours."

Infrared lamps used in the home and for therapeutic purposes are commonly of the convenient self-contained 250-watt R-10 bulb type with internal reflector and red bulb. Those used in industrial processes are of three types: reflector lamps (125-, 250-, and 375-watt R-40), clear G-30 bulb lamps (125, 250, 375 and 500 watts), and the more recently developed small linear sources in the T-3 quartz bulb. The latter are available in a number of sizes, and the effective heating length and the voltage rating increases with the wattage. Gold-plated or specular aluminum reflectors are most effective for use with unreflectorized infrared lamps.

UNIT 6

Mercury Lamps

A special type of lighting unit which has become very popular for use in industrial plants and outdoor lighting applications is the mercury-vapor lamp. This type of lamp produces light directly as a result of a current passed through gas or vapor under pressure.

While a lighted mercury lamp appears to emit a white light, it actually produces light with a predominance of yellow and green rays and a small percentage of violet and blue rays. Red is absent, and therefore red objects appear black or dark brown under mercury-vapor lamps. This color distortion has, in the past, prevented its use for many applications, but has now been overcome to a certain extent by the use of red-light-generating chemicals within the bulb. Mercury lamps are presently being used indoors for more and more commercial lighting applications.

Typical mercury lamp parts are shown in Fig. 6-1. The basic elements are: an arc tube, made of quartz to withstand the high temperatures resulting when the lamp builds up to normal wattage; two main operating electrodes, located at opposite ends of the tube; a starting electrode in series with a starting resistor and connected to the lead wire of the lower operating electrode; tube leads and supports; and an outer phosphor-coated glass bulb that helps to stabilize the lamp operation and prevents oxidation of metal parts.

Designation of Mercury Lamps

The American National Standards Institute has developed a system of codes for mercury lamp types. This system indicates a letter "H" followed by a number and two letters. The letter "H" stands for the chemical symbol "Hg" for mercury and indicates the lamp is a mercury type. The number represents the ballast type, and the two following letters define the physical lamp characteristics. Additional letters are used to identify the type of phosphor coating on the inside of the bulb. They are as follows:

C—Color-improved phosphor.
W—High-efficiency phosphor.
DX—Deluxe.
Y—Yellow.

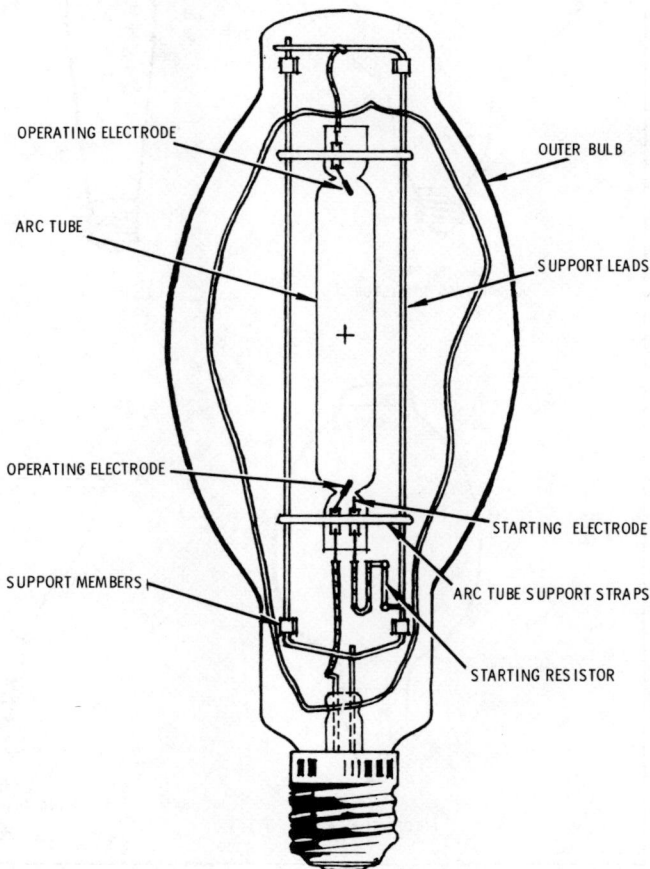

Fig. 6-1. A phosphor-coated mercury lamp.

Bulb Shapes

As shown in Fig. 6-2, mercury lamps are manufactured in a variety of bulb shapes for various lighting applications. All of these mercury lamps are provided with conventional medium or mogul screw bases.

Lamp Types

Mercury lamps for general lighting purposes are available in wattages from 40 to 2000 watts, in both clear and phosphor-coated bulbs. Any type of mercury lamp requires its own specially designed transformer, or ballast, for proper starting and operating performance. These ballasts are usually located externally from the lamp. However, self-ballasted lamps with built-in filament-type ballasts are now available.

Operating Characteristics

Long Life—Mercury lamps have inherently long life and may last over 24,000 hours. However, during such long life, the light output decreases to less than half the initial rated output, and it becomes expensive to keep an old lamp in service.

Voltage—Mercury lamps must be operated within rather close limits of the proper voltage for proper operation. Undervoltage reduces starting reliability; excessive voltage causes increased wattage, which in turn is likely to raise both lamp and transformer temperatures beyond safe limits. Most transformers have several top connections so that the one corresponding closely to the external line voltage may be used.

Starting—Starting the mercury lamp takes a few cycles—when voltage is first applied, a light

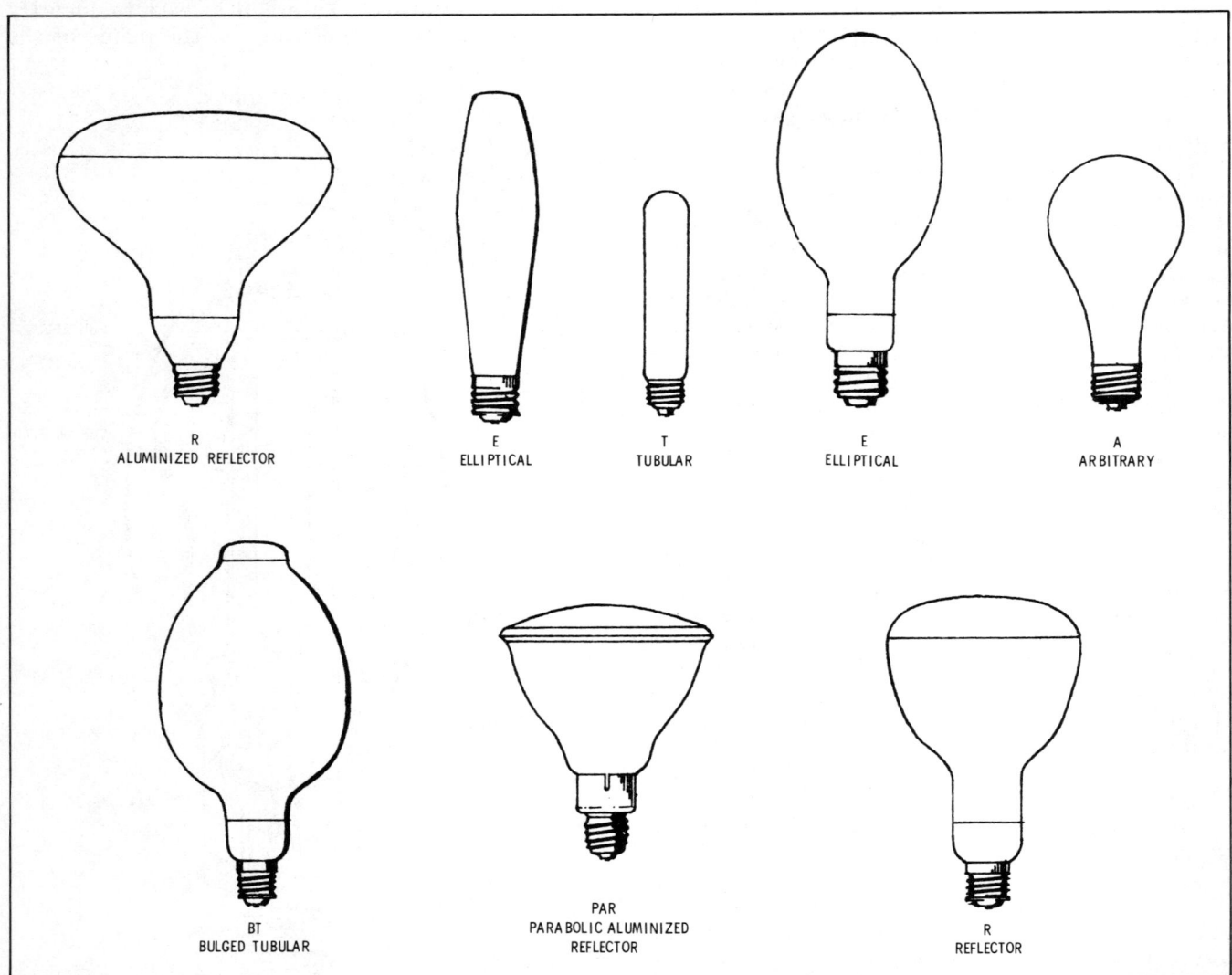

Fig. 6-2. Some typical mercury lamp bulb shapes.

bluish glow fills the arc tube between the starting and operating electrode. As the mercury gradually vaporizes, the lamp voltage increases rapidly, and the current decreases somewhat until all the mercury is vaporized and the temperature and vapor pressure become stabilized to a point where an arc is established between main electrodes.

Warm-up—It usually requires four to seven minutes for all the mercury in a lamp to vaporize and for the lamp to come up to full brightness. At this time the voltage and current are stable; however, if the current is interrupted—even momentarily—the lamp may be extinguished and cannot be relighted again until it has cooled enough to reduce the mercury-vapor pressure sufficiently to allow the arc to go through the starting cycle again.

Restarting—Mercury lamps are extinguished in case of current interruption or excessive low voltage. They will not restart until they have cooled down and the internal vapor pressure has been reduced to the point of restarting the arc with the voltage available. The restarting-time interval is between four to seven minutes.

The long restarting time is one disadvantage of mercury lamps for some applications. It is recommended that mercury lamps be combined with an auxiliary system of incandescent or fluorescent lamps when used in areas where an accidental current interruption may extinguish the mercury lamps and cause accidents or panic among people with no other source of light.

Efficiency—Current mercury lamps produce from 30 to 65 lumens per watt depending on wattage and color.

Normal Operation—The output of mercury lamps is not affected by ambient temperature, and this type of lamp is therefore suited for outdoor applications. However, low temperatures or high winds may produce a condition where higher voltage is needed to start the lamp.

ADDITIONAL CHARACTERISTICS

In addition to being used to produce visible light, mercury lamps have been used to produce ultraviolet radiation as a source of black light. This type of light requires that all visible light be screened out by filters that allow only the ultraviolet energy to pass through. These screening filters are made of special dark-blue or red-purple glass.

Black-light lamps have widely diverse applications, such as decorative and theatrical effects, industrial inspection, medical and chemical analysis, criminal investigations, mineralogy, and varied military applications.

METAL-HALIDE LAMPS

The metal-halide lamps closely resemble a regular clear mercury lamp, but the inner arc tube contains additional halide chemical compounds to increase the light output and improve the lamp color. Fig. 6-3 illustrates the basic components of a metal-halide lamp.

APPLICATION

Since the color produced by metal-halide lamps is much "warmer" than regular mercury lamps, it is suitable for many indoor applications including food displays. It has found more use, however, in outdoor floodlighting, sports-lighting, and certain general street-lighting applications.

The efficiency of the metal-halide lamp is approximately twice that of conventional mercury

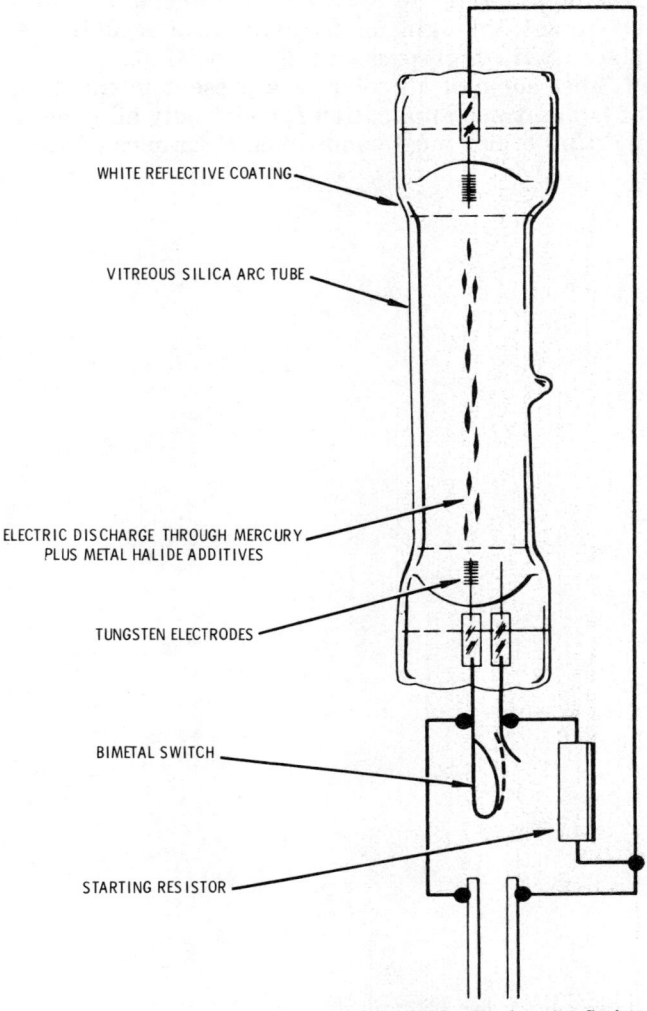

Courtesy Illuminating Engineering Society

Fig. 6-3. Basic components of a metal-halide lamp.

lamps—producing from 75 to 105 lumens per watt. However, the life of this lamp is shorter than regular mercury lamps, which average from 6,000 to 10,000 hours. All other operating characteristics are similar to those of the regular mercury lamp described earlier in this unit.

HIGH-PRESSURE SODIUM LAMPS

The high-pressure sodium lamp operates in a similar fashion to other discharge lamps in that it utilizes an arc tube to enclose gases through which an electric current passes. However, the unique light-transmitting ceramic tube enables sodium to be operated at higher temperatures and pressures than were previously attainable. The result is a warm yellow light at nearly maximum theoretical efficiency—100 to 115 lumens per watt.

This lamp is excellent for street lighting and general outdoor area lighting. After an extensive feasibility study in Baltimore, Maryland, nearly all downtown lighting fixtures were recently replaced with high-pressure sodium lamps.

Since some of all colors are present in this type of lamp, it has application for virtually all general lighting under most conditions. Examples of some applications of this lamp will be discussed in later units.

ADVANTAGES AND DISADVANTAGES OF HIGH-INTENSITY DISCHARGE (MERCURY) LAMPS

Advantages
- Long life.
- High lumen output.
- Have compactness of incandescent lamps.
- Not affected by ambient temperatures as are fluorescent lamps.
- Better degree of light control than with fluorescent lamps.

Disadvantages
- Color acceptability low with clear mercury.
- Sensitive to voltage variation.
- Long restarting time required.
- Light is extinguished if momentary current interruption occurs.
- Delay of 4 to 7 minutes from starting time to full brightness.

UNIT 7

Fluorescent Lamps

Fluorescent lamps have become the major light source for general interior lighting of commercial and institutional buildings and have challenged other sources for residential, exterior, and other lighting applications.

As shown in Fig. 7-1, the fluorescent lamp is available in straight, U-shaped, or circular configurations, and in various diameters.

A fluorescent lamp consists of an airtight glass tube enclosing a small drop of mercury and a small amount of argon or argon-neon gas to facilitate starting the arc. After the arc is started, the mercury vapor emits ultraviolet radiation which is invisible and does not pass through the glass. However, the inside of the glass tube is coated with a highly sensitive fluorescent powder (phosphors) which is activated by the ultraviolet radiation and in turn converts the invisible energy to visible light. By mixtures of various phosphors, a wide range of visible light colors is possible.

COLORS OF FLUORESCENT LAMPS

Cool White

This lamp is often selected for offices, factories, and commercial areas where a psychologically cool working atmosphere is desirable. This is the most popular of all fluorescent lamp colors since it gives a natural outdoor lighting effect and is one of the most efficient fluorescent lamps manufactured today.

Deluxe Cool White

This lamp is used for the same general applications as the cool white, but contains more red which emphasizes pink skin tones and is therefore more flattering to the appearance of people. Deluxe cool white is also used in food display because it gives a good appearance to lean meat; keeps fats looking white; and emphasizes fresh, crisp appearance of green vegetables. This type of lamp is generally chosen wherever very uniform color rendition is desired, although it is less efficient than cool white.

Warm White

Warm-white lamps are used whenever a warm social atmosphere is desirable in areas that are not color critical. It approaches incandescent in color and is suggested whenever a mixture of fluorescent and incandescent lamps is used. While it gives an acceptable appearance to people, it has some tendency to emphasize sallowness. Yellow, orange, and tan interior finishes are emphasized by this lamp, and its beige tint gives a bright warm appearance to reds; brings out the yellow in green; and adds a warm tone to blue. It imparts a yellowish white or yellowish gray appearance to neutral surfaces.

Deluxe Warm White

Deluxe warm-white lamps are more flattering to complexions than warm white and are very similar to incandescent lamps in that they impart a ruddy or tanned hue to the skin. It is generally recommended for home or social environment applications and for commercial use where flattering effects on people and merchandise are considered important. This type of lamp enhances the appearance of poultry, cheese, and baked goods. These lamps are approximately 25% less efficient than warm-white lamps.

White

White lamps are used for general lighting applications in offices, schools, stores, and homes where

PRINCIPLES OF ILLUMINATION

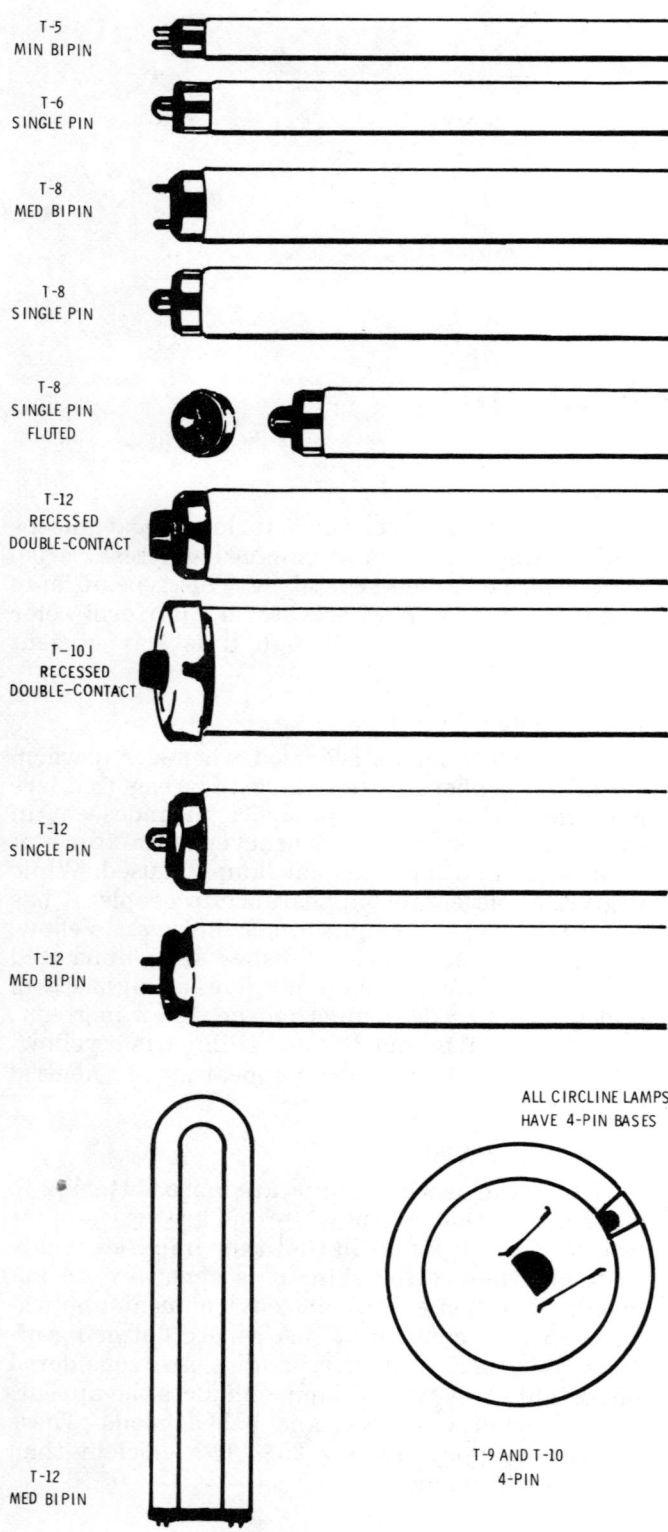

Fig. 7-1. An illustration of different sizes and shapes of fluorescent lamps.

either a cool working atmosphere or warm social atmosphere is not critical. They emphasize yellow, yellow green, and orange interior finishes.

Daylight

Daylight lamps are for use in industry and work areas where the blue color associated with the "north light" of actual daylight is preferred. While it makes blue and green bright and clear, it tends to tone down red, orange, and yellow.

In general the designations "warm" and "cool" represent the differences between artificial light and natural daylight in the appearance they give to an area. Their deluxe counterparts have a greater amount of red light, supplied by a second phosphor within the tube. The red light shows colors more naturally, but at a sacrifice in efficiency.

Other colors of fluorescent lamps are available in sizes that are interchangeable with white lamps. These colored lamps are best used for flooding large areas with colored light; where a colored light of small area must be projected at a distant object, incandescent lamps using colored filters are best.

CLASSES OF FLUORESCENT LAMPS

Preheat Lamps

Preheat hot-cathode fluorescent lamps use a two-pin base and a starter which provides momentary current flow through the filament cathode in order to heat them. The radiation from the cathodes is possible only after the cathodes have been preheated. The time interval necessary for preheating is one drawback of this type of lamp, but this drawback is offset by the significant savings in ballast design and lamp life. The preheat-type diagram using a single-lamp ballast with a capacitor is shown in Fig. 7-2. The switch and starter connected across the lamp can be either automatic or manual.

Instant Start

In order to overcome the slow starting of the preheat system and eliminate the need for a starter, the instant-start lamp was developed. Instant starting is accomplished by use of a specially designed ballast which delivers a high starting voltage and a normal operating voltage once the lamps are started. Because no preheating is necessary with instant-start lamps, only a single pin on each end of the lamp is required. Hot-cathode lamps with single-pin bases are called slimline lamps. The wiring diagram of a circuit with two bipin instant-start lamps with a ballast is shown in Fig. 7-3.

FLUORESCENT LAMPS

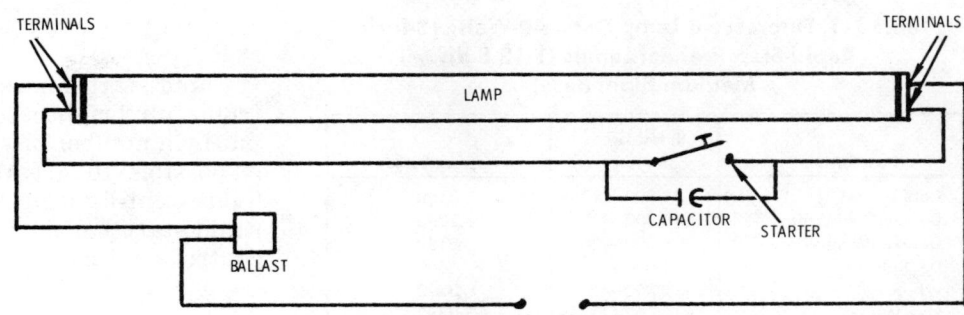

Fig. 7-2. A preheat fluorescent lamp diagram.

Rapid-Start Lamps

A rapid-start fluorescent lamp retains the advantage of preheat starting, speeds up the starting interval, and eliminates the separate starter switch. The smooth, rapid start is accomplished by a built-in electrode heating coil in the ballast, and the lamp lights almost as quickly as instant-start lamps. These lamps are the most popular and important for use in fluorescent-lighting systems.

Today, the most commonly used lamp is a rapid-start lamp operating at 430 milliamperes, or approximately 10 watts per foot of lamp.

Table 7-1 lists the characteristics of rapid-start lamps which do not require a separate starter.

ALL-WEATHER FLUORESCENT LAMPS

Fluorescent lamps with common base sizes operate most efficiently at normal room temperatures of 70 to 80°F, at which the temperature of the glass tube itself is between 100 and 120°F. Where temperatures fall below this level, as in outdoor applications during winter months, a jacket placed around the outside of the lamp will maintain bulb wall temperature and will help provide reasonable light output. Rapid-start lamps with this jacket are known as all-weather fluorescent lamps.

COLD-CATHODE FLUORESCENT LAMPS

Cold-cathode-type circuits have been used for years in neon-sign tubing because it operates at relatively low current in small-diameter tubing adaptable to bending into sign letters or luminous patterns. All lamps are instant start and require special high-voltage circuits. For the same bulb size, phosphor and current loading, the lumen output and maintenance of cold-cathode lamps are identical in performance to those with a hot cathode. These types of lamps find greatest use in sign and display lighting. For some general applications, there are standardized lengths of tubing produced in T-8 glass envelopes and four, six, and eight feet in length, but few such tubes are being installed at present.

BALLAST

Every fluorescent lamp needs a ballast in order to operate. The ballast is simply a coil of insulated wire wound on a frame, or core, made up of thin layers of iron stampings, and it performs any or all of the following functions:

1. Limits the current flow through the lamp to the value for which the lamp is designed.

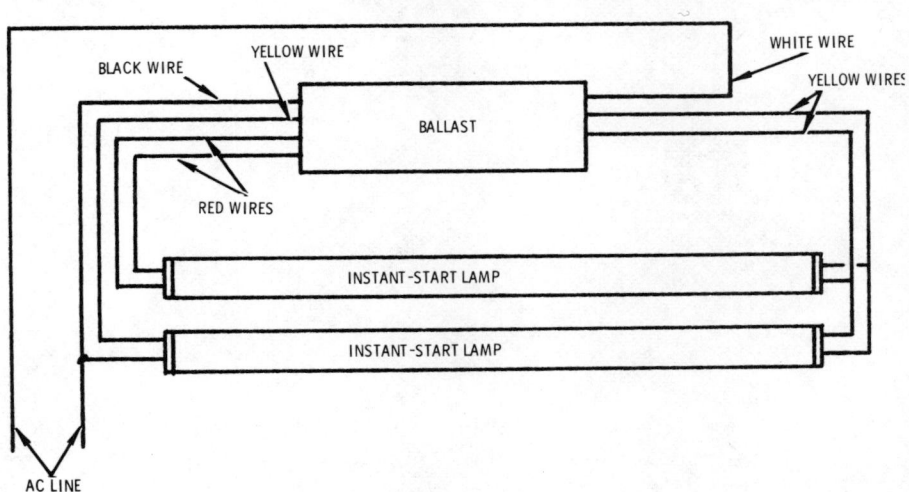

Fig. 7-3. A wiring diagram for two bipin instant-start lamps with a ballast.

PRINCIPLES OF ILLUMINATION

Table 7-1. Fluorescent Lamp Data, 40-Watt, 48-Inch, Rapid-Start Preheat Lamps (T-12 Bulb, Medium Bipin Base)

Description	Rated Life (hours)	Approx Lumen
Cool White	18,000+	3150
Deluxe Cool White	18,000+	2200
Deluxe Warm White	18,000+	2150
Daylight	18,000+	2600
White	18,000+	3200
Sign White	18,000+	2440
Warm White	18,000+	3200
Chroma 55	18,000+	2020
Chroma 75	18,000+	1990
Soft White	18,000+	1990
Cool White	15,000	3250
Daylight	15,000	2650
White	15,000	3300
Warm White	15,000	3300
Cool Green	18,000+	2650
Plant Light	12,000	900
Blue	12,000	1160
Green	12,000	4500
Gold	12,000	2400
Pink	12,000	1160
Red	12,000	200
Cool White Reflector 135° Window	18,000+	2600
Warm White Reflector 135° Window	18,000+	2650
*Cool White	12,000	2800
*Warm White	12,000	2800
†Black Light	12,000	
†Black Light Blue	12,000	

*U-shape bulb 22½ inches long
†Black light

2. Causes a drop in line voltage and provides the desired lamp voltage, which in turn determines the rated current in the lamp.
3. Provides power factor correction.
4. Provides radio interference suppression.

DIMMING

Rapid-start fluorescent lamps can be dimmed from full brightness to approximately zero output by a number of special circuits. This has made it possible to greatly increase the flexibility of fluorescent-lighting systems. However, a special rapid-start ballast in conjunction with dimming-control devices is required.

FLASHING

Rapid-start and cold-cathode lamps can be flashed without any appreciable loss in lamp life, since a special ballast can be used to provide continuous electrode heating while the current to the arc is interrupted. This technique will be fully covered in Units 19 and 25.

ADVANTAGES AND DISADVANTAGES OF FLUORESCENT LAMPS

Advantages
- High efficiency.
- Long life.
- A linear source of light.
- Variety of colors available.
- Relatively low surface brightness.
- Economy in operation.

Disadvantages
- Very sensitive to temperature and humidity.
- Radio interference.
- Difficult to control.
- Higher initial installation cost.

SECTION III

PRINCIPLES OF LIGHTING DESIGN

UNIT 8

Basic Design Procedure

The basic requirement for any lighting design is to determine the amount of light that should be provided and the best means of providing it. However, since individual tastes and opinions vary greatly, there can be many suitable solutions to the same lighting problem. Some of these solutions will be dull and commonplace, while others will show imagination and resourcefulness. The lighting designer should always strive to select lighting equipment that will provide the highest visual comfort and performance that is consistent with the type of area to be lighted and the budget provided.

SURVEY

A survey of a particular lighting problem is necessary, prior to actual calculations, in order to determine the best possible lighting system and the best method of providing that system. The survey is usually most effective when it follows a specific and systematic pattern. The collection of necessary data for this survey may be made quickly and efficiently by using a prepared form such as the one illustrated in Fig. 8-1. Instructions for using this form are given in Units 8 and 9.

THE LIGHTING SURVEY

The lighting designer should insert the date, his initials, the name of the project, and the room or area designation in the spaces provided on the form. The following discussion will be to provide an appropriate lighting system for the warehouse building in Fig. 8-2.

Recommended illumination levels for various areas can be found in Appendix A. Looking under Storage Rooms and Warehouses in Appendix A, we find that the recommended illumination level for a medium-active warehouse is 20 footcandles. (It has been determined that the activity in this particular warehouse will be medium.) Insert 20 footcandles in the appropriate space on the form.

Room Dimension

A complete description of the physical characteristics of the room is necessary and should include: length, width, ceiling height, and total floor area. This physical data is obtained from blueprints or actual measurements. Inside dimensions should be used. The floor plan, in Fig. 8-2, gives the interior length of the warehouse as 80 feet, the width as 60 feet, and the ceiling height as 12 feet; thus, the total area is (80 × 60) 4800 square feet. Enter this data in the appropriate space on the form.

Surface Reflectance

Surface reflectance is the percentage of the total amount of light which is reflected from the surface. In general, light-colored surfaces will have a higher percentage of reflectance than those with dark finishes. We will assume that the warehouse walls will be unpainted concrete block; the ceiling a dark-colored surface; and the floor unfinished concrete. By comparing color samples of known reflections with our warehouse surfaces, we are able to determine the reflectances of the warehouse surfaces as follows: ceiling 50%; walls, 30%; and floor, 20%. Enter these percentages in the appropriate spaces on the form.

Cavity

The zonal-cavity system divides the room into three cavities as shown in Fig. 8-3. These dimensions should be entered in the appropriate spaces provided on the form. Referring to Fig. 8-3,

PRINCIPLES OF ILLUMINATION

LIGHTING SURVEY FORM

Date __3-3-73__   Designed By __JEB__

Project Name __ABC Warehouse__   Room or Area __Storage__

Recommended Illumination Level __20 fc__   Height __12 ft__

A. ROOM DIMEN: Length __80 ft__   Width __60 ft__   Area __4800 Sq ft__

B. SURFACE REFLECT: Ceiling __50%__   Wall __30%__   Floor __20%__

   Fixture Mounting Height __11 ft__

C. CAVITY DATA   Room Cavity: Height __8 ft__   Ratio _____

   Ceiling Cavity: Height __1 ft__   Ratio _____   Eff. Reflectance _____

   Floor Cavity: Height __3 ft__   Ratio _____   Eff. Reflectance _____

D. FIXTURE DATA

   Mfr. __Goodylite__   Cat. No. __333__

   Lamps per Fixture __1-200 W IF__   Lumens per Lamp __4010__

   Coeff. of Utilization _____   Maintenance Factor _____

Fig. 8-1. A typical lighting survey form.

---

we find that the ceiling cavity is one foot; the room cavity is eight feet; and the floor cavity is three feet. If recessed or shallow surface-mounted lighting fixtures were used, the ceiling cavity would be zero. However, in this case the light source is suspended approximately one foot from the ceiling, so the ceiling cavity height is one foot.

The desk top is approximately three feet above the finished floor. Thus, the working plane is three feet above the finished floor; therefore, the floor cavity is three feet.

The remaining area between the ceiling and floor cavity is called the room cavity, and in this case it is eight feet.

Fixture

The advantages and use of various types of lamps were covered in Section II along with their related luminaires. In the case of the warehouse illustrated in Fig. 8-2, it has been decided to use industrial-type reflector lighting fixtures with incandescent lamps such as illustrated in Fig. 8-4.

This particular type of luminaire was chosen for the following reasons:

1. Low initial cost.
2. It will operate over a wide range of ambient

BASIC DESIGN PROCEDURE

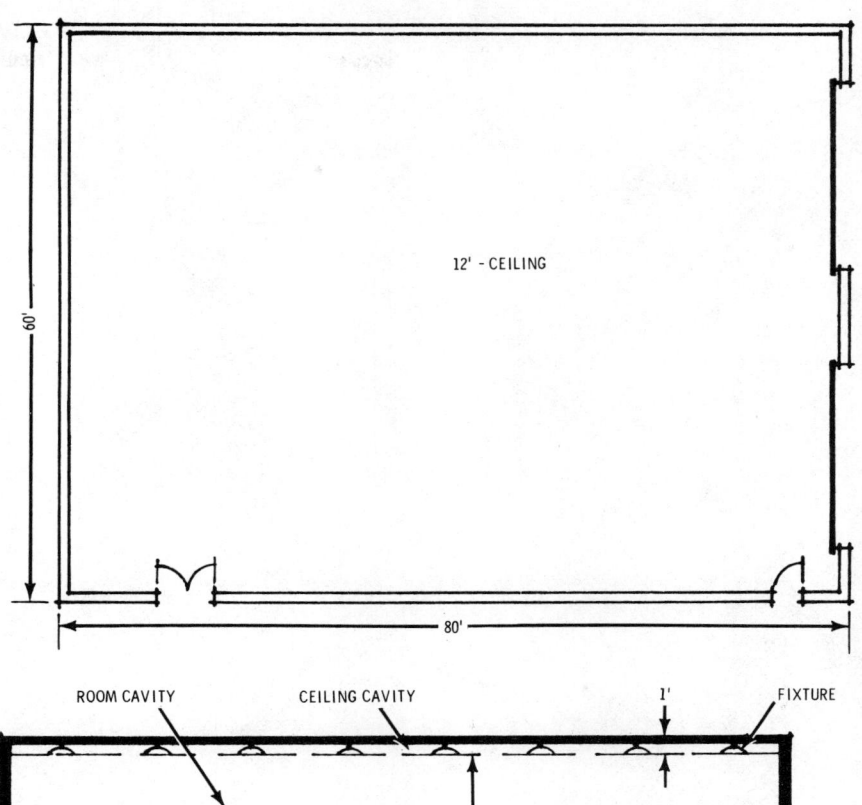

Fig. 8-2. Warehouse for which lighting system is to be designed.

Fig. 8-3. A zonal-cavity system.

temperatures, such as will be encountered in the warehouse.

3. Since the footcandle level is relatively low, the higher operating temperature of the few incandescent fixtures will not seriously affect the room temperature.

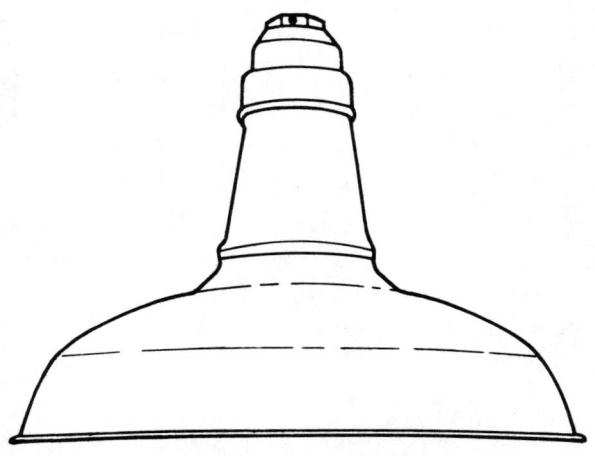

Fig. 8-4. An industrial-type reflector lighting fixture.

We will call this light a Model 333 manufactured by the Goodylite Corporation. Enter this data in the appropriate spaces on the form.

Each fixture will contain one 200-watt inside-frosted lamp. Referring to lamp data tables, we find that a 200-watt inside-frosted lamp gives off approximately 4010 initial lumens. This data is also entered in the appropriate spaces on the form.

SUMMARY

A lighting survey is necessary in order to determine the best possible lighting system and the best method of providing that system.

A survey usually includes:

1. The physical dimensions of the room or area.
2. The room cavity data.
3. The selection of the most suitable light source.
4. The number and size of lamps per fixture.
5. The lumens per lamp.

47

UNIT 9

Lumen Method Zonal-Cavity System

This unit describes a fundamental approach to designing a lighting system. It also introduces two general applications of the zonal cavity method of lighting calculations: first to determine how many lighting fixtures, or luminaires, are required to produce a given lighting level in footcandles; second, to determine what lighting level will be produced by a given number of fixtures.

ZONAL-CAVITY METHOD

The zonal-cavity method of calculating average illumination levels assumes each room or area to consist of the following three separate cavities: ceiling cavity, room cavity, and floor cavity.

Fig. 9-1 shows that the *ceiling cavity* extends from the lighting fixture plane upward to the ceiling. The *floor cavity* extends from the work plane downward to the floor, while the *room cavity* is the space between the lighting fixture plane and the working plane.

If the lighting fixtures are recessed or surface-mounted on the ceiling, there will be no ceiling cavity and the ceiling-cavity reflectance will be equal to the actual ceiling reflectance. Similarly, if the work plane is at floor level, there will be no floor cavity and the floor-cavity reflectance will be equal to the actual floor reflectance. The geometric proportions of these spaces become the "cavity ratios."

CAVITY RATIO

Rooms are classified according to shape by *ten* cavity-ratio numbers. The basic formula for obtaining cavity ratios in rectangular-shaped rooms is:

$$\text{Cavity Ratio} = \frac{5 \times \text{Height (Length + Width)}}{\text{Length} \times \text{Width}}$$

where height is the height of the cavity under consideration—that is, ceiling, floor, or room cavity.

For example, assume the room illustrated in Fig. 9-1 is 8 feet wide by 12 feet in length. The lighting fixtures are suspended 1 foot below the ceiling. Find the ceiling cavity ratio. By substituting known values in the previous formula we have:

$$\text{Ceiling Cavity Ratio} = \frac{5 \times 1\ (12 + 8)}{12 \times 8}$$
$$= 1.04 \text{ or } 1$$

For rooms composed of more than one rectangle, such as an L-shaped room, the cavity ratio is obtained by the following formula:

$$\text{Cavity Ratio} = \frac{2.5 \times \text{Wall Area}}{\text{Floor Area}}$$

In calculating the ceiling cavity ratio, wall area is determined by multiplying the total linear feet of the walls by the distance between the lighting fixture plane and the ceiling cavity.

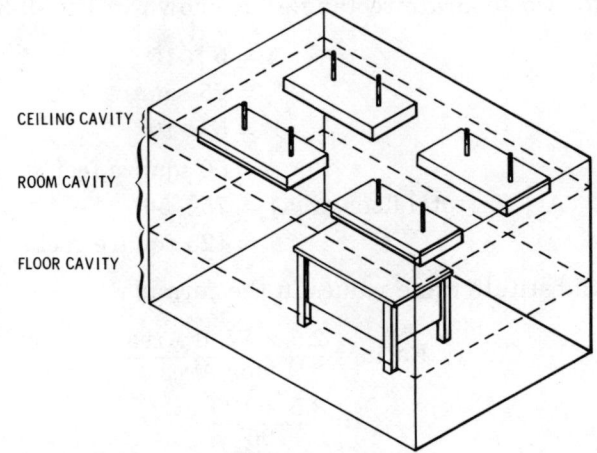

Fig. 9-1. The three room cavities.

PRINCIPLES OF ILLUMINATION

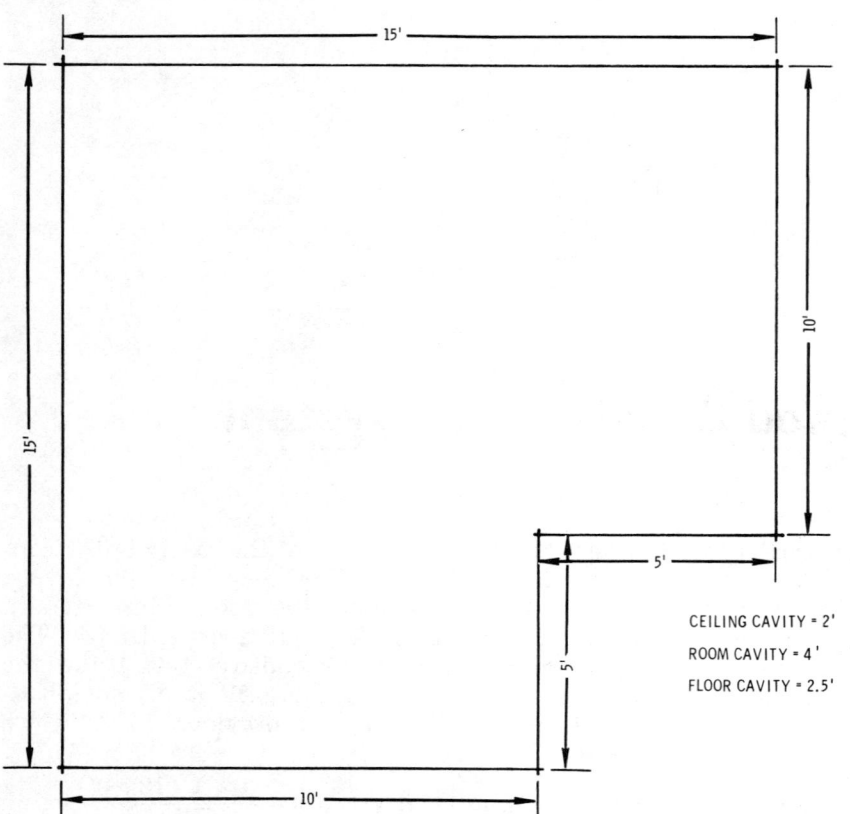

Fig. 9-2. Lighting fixture suspended below ceiling in L-shaped room.

CEILING CAVITY = 2'
ROOM CAVITY = 4'
FLOOR CAVITY = 2.5'

For example, an L-shaped room has the physical dimensions as illustrated in Fig. 9-2. Notice that the ceiling cavity is two feet deep. Find the ceiling cavity ratio of this room.

Find the total linear feet of the walls: $15 + 15 + 10 + 10 + 5 + 5 = 60$ linear feet.

Multiply the total linear feet by the ceiling cavity depth: $60 \times 2 = 120$ square feet.

Find the total floor area by dividing the room into two separate rectangles as shown in Fig. 9-3.

$$A = 5 \times 15$$
$$= 75 \text{ square feet}$$
$$B = 5 \times 10$$
$$= 50 \text{ square feet}$$
$$A + B \text{ (total floor area)} = 75 + 50$$
$$= 125 \text{ square feet}$$

Substitute these values in the formula.

$$\text{Cavity Ratio} = \frac{2.5 \times \text{Wall Area}}{\text{Floor Area}}$$
$$= \frac{2.5 \times 120}{125}$$
$$= 2.4$$

Similarly, the floor cavity height in Fig. 9-2 is 2.5 feet. Since we already know the perimeter of the room to be 60 feet, we can multiply 2.5 by 60 and obtain a floor-cavity wall area of 150 square feet. The floor area remains the same for all three cavity calculations. Thus,

$$\text{Floor Cavity Ratio} = \frac{2.5 \times 150}{125}$$
$$= 3$$

For calculating the room cavity ratio, the wall area is determined by multiplying the total linear feet of the wall (60 feet in this case) by the height of the room cavity. Again refer to Fig. 9-2, which shows the height of the room cavity to be 4 feet. The room cavity wall area is then (4×60) 240 square feet. The floor area has previously been determined as 125 square feet. Thus,

$$\text{Room Cavity Ratio} = \frac{2.5 \times 240}{125}$$
$$= 4.8$$

For other than rectangular rooms, the area can be calculated as required. For example, in a circular room, the cavity wall area equals height $\times$ $2\pi r$ and the floor area equals πr^2. Thus,

LUMEN METHOD-ZONAL-CAVITY SYSTEM

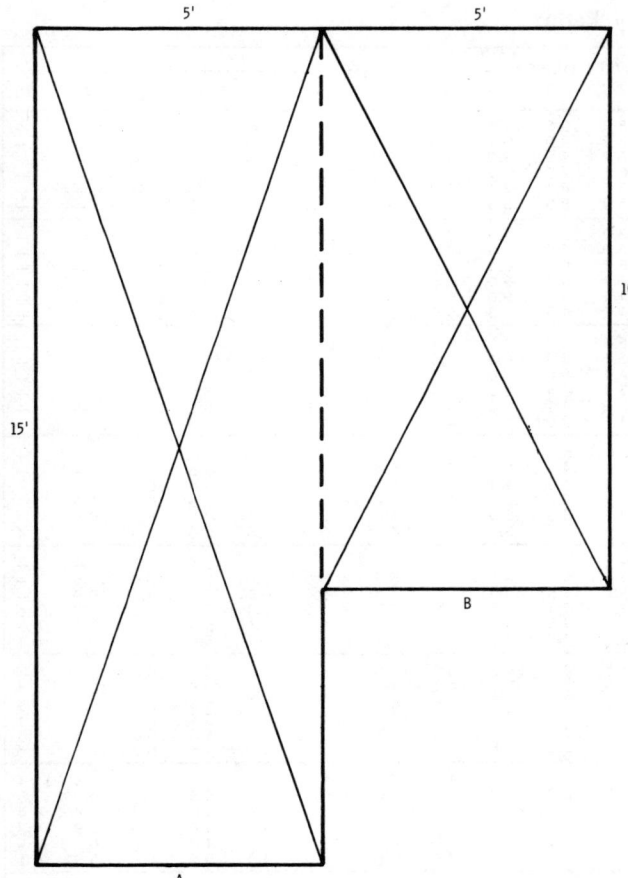

Fig. 9-3. Dividing the L-shaped room into two separate rectangles.

$$\text{Cavity Ratio} = \frac{2.5 \times \text{height} \times 2\pi r}{\pi r^2} \text{ or } \frac{5 \text{ height}}{r}$$

Table 9-1 may be used to determine the cavity ratios for rectangular rooms. Using this table is usually faster than working the formulas and is usually preferred by lighting designers. Find the width of the room under consideration in the column on the far left. Then go to the next column and find the length of the room. Continue across this line until a number lines up with the column of the cavity depth (found at top of page). This number is the cavity ratio.

EFFECTIVE REFLECTANCE

Before the coefficient of utilization can be selected, the combination of ceiling and wall reflectance as well as floor and wall reflectance must be converted to *effective ceiling or floor reflectance*. The effective reflectance of the ceiling and floor cavities takes into account the effect of inter-reflection of light among the various room surfaces. Again, charts or tables are provided for this conversion (Table 9-2). In order to find the effective reflectance, locate the column containing the known percentage of ceiling or floor reflectance and wall reflectance. Then continue down this column and read the effective reflectance opposite the appropriate cavity ratio.

As an example, assume the room illustrated in Fig. 9-1 has a ceiling reflectance of 80% and a wall reflectance of 50%. We have previously calculated the ceiling cavity ratio (1). By following the instructions given in the preceding paragraphs, we find that the effective ceiling reflectance is 66%.

COEFFICIENT OF UTILIZATION

Manufacturers usually supply photometric data for their own lighting fixtures. This data contains coefficients of utilization for various surface reflections and room cavity ratios. Appendix B also lists photometric data for several typical lighting fixtures and can be used to solve every problem found in this text. However, a more accurate calculation will result in actual applications, if the coefficients of utilization of the actual lighting fixture under consideration are used. Table 9-3 is typical of such data. This table of coefficients is based on an effective floor-cavity reflectance of 20%. If it is substantially different, an adjustment is made later.

Assume that we want the coefficient of utilization for: effective ceiling-cavity reflectance (% CCR) of 0.78, wall reflectance (% WR) of 0.5 and room cavity ratio (RCR) of 4 from Table 9-3. The effective floor-cavity reflectance (% FCR) is 18% and is close enough to 20% for our purposes. The 78% ceiling reflectance is also close enough to the 80% reflectance in Table 9-3 without adjustment.

For an 80% effective ceiling-cavity reflectance, a 50% wall reflectance, and a room cavity ratio of 4, Table 9-3 gives a coefficient of utilization of 0.43.

CORRECTION FACTOR AND INTERPOLATION

If the effective floor-cavity reflectance is 18%, 19%, 21%, or 22%, direct use of coefficient of utilization tables can be made. However, if the effective floor-cavity reflectance is 17% or below, or 23% and above, the lighting designer may find that an adjustment is necessary. Table 9-4 lists correction factors for effective floor-cavity reflectances other than 20%.

In order to demonstrate the use of Table 9-4 and to provide an example of interpolation, assume the effective floor reflectance is 17%.

Table 9-1. Cavity Ratios

ROOM DIMENSIONS		\multicolumn{19}{c}{CAVITY DEPTH}
Width	Length	1.0	1.5	2.0	2.5	3.0	3.5	4.0	5.0	6.0	7.0	8	9	10	11	12	14	16	20	25	30
8	8	1.2	1.9	2.5	3.1	3.7	4.4	5.0	6.2	7.5	8.8	10.0	11.2	12.5	—	—	—	—	—	—	—
	10	1.1	1.7	2.2	2.8	3.4	3.9	4.5	5.6	6.7	7.9	9.0	10.1	11.3	12.4	—	—	—	—	—	—
	14	1.0	1.5	2.0	2.5	3.0	3.4	3.9	4.9	5.9	6.9	7.8	8.8	9.7	10.7	11.7	—	—	—	—	—
	20	0.9	1.3	1.7	2.2	2.6	3.1	3.5	4.4	5.2	6.1	7.0	7.9	8.8	9.6	10.5	12.2	—	—	—	—
	30	0.8	1.2	1.6	2.0	2.4	2.8	3.2	4.0	4.7	5.5	6.3	7.1	7.9	8.7	9.5	11.0	—	—	—	—
	40	0.7	1.1	1.5	1.9	2.3	2.6	3.0	3.7	4.5	5.3	5.9	6.5	7.4	8.1	8.8	10.3	11.8	—	—	—
10	10	1.0	1.5	2.0	2.5	3.0	3.5	4.0	5.0	6.0	7.0	8.0	9.0	10.0	11.0	12.0	—	—	—	—	—
	14	0.9	1.3	1.7	2.1	2.6	3.0	3.4	4.3	5.1	6.0	6.9	7.8	8.6	9.5	10.4	12.0	—	—	—	—
	20	0.7	1.1	1.5	1.9	2.3	2.6	3.0	3.7	4.5	5.3	6.0	6.8	7.5	8.3	9.0	10.5	12.0	—	—	—
	30	0.7	1.0	1.3	1.7	2.0	2.3	2.7	3.3	4.0	4.7	5.3	6.0	6.6	7.3	8.0	9.4	10.6	—	—	—
	40	0.6	0.9	1.2	1.6	1.9	2.2	2.5	3.1	3.7	4.4	5.0	5.6	6.2	6.9	7.5	8.7	10.0	12.5	—	—
	60	0.6	0.9	1.2	1.5	1.7	2.0	2.3	2.9	3.5	4.1	4.7	5.3	5.9	6.5	7.1	8.2	9.4	11.7	—	—
12	12	0.8	1.2	1.7	2.1	2.5	2.9	3.3	4.2	5.0	5.8	6.7	7.5	8.4	9.2	10.0	11.7	—	—	—	—
	16	0.7	1.1	1.5	1.8	2.2	2.5	2.9	3.6	4.4	5.1	5.8	6.5	7.2	8.0	8.7	10.2	11.6	—	—	—
	24	0.6	0.9	1.2	1.6	1.9	2.2	2.5	3.1	3.7	4.4	5.0	5.6	6.2	6.9	7.5	8.7	10.0	12.5	—	—
	36	0.6	0.8	1.1	1.4	1.7	1.9	2.2	2.8	3.3	3.9	4.4	5.0	5.5	6.0	6.6	7.8	8.8	11.0	—	—
	50	0.5	0.8	1.0	1.3	1.5	1.8	2.1	2.6	3.1	3.6	4.1	4.6	5.1	5.6	6.2	7.2	8.2	10.2	—	—
	70	0.5	0.7	1.0	1.2	1.5	1.7	2.0	2.4	2.9	3.4	3.9	4.4	4.9	5.4	5.8	6.8	7.8	9.7	12.2	—
14	14	0.7	1.1	1.4	1.8	2.1	2.5	2.9	3.6	4.3	5.0	5.7	6.4	7.1	7.8	8.5	10.0	11.4	—	—	—
	20	0.6	0.9	1.2	1.5	1.8	2.1	2.4	3.0	3.6	4.2	4.9	5.5	6.1	6.7	7.3	8.6	9.8	12.3	—	—
	30	0.5	0.8	1.0	1.3	1.6	1.8	2.1	2.6	3.1	3.7	4.2	4.7	5.2	5.8	6.3	7.3	8.4	10.5	—	—
	42	0.5	0.7	1.0	1.2	1.4	1.7	1.9	2.4	2.9	3.3	3.8	4.3	4.7	5.2	5.7	6.7	7.6	9.5	11.9	—
	60	0.4	0.7	0.9	1.1	1.3	1.5	1.8	2.2	2.6	3.1	3.5	3.9	4.4	4.8	5.2	6.1	7.0	8.8	10.9	—
	90	0.4	0.6	0.8	1.0	1.2	1.4	1.6	2.0	2.5	2.9	3.3	3.7	4.1	4.5	5.0	5.8	6.6	8.3	10.3	12.4
17	17	0.6	0.9	1.2	1.5	1.8	2.1	2.3	2.9	3.5	4.1	4.7	5.3	5.9	6.5	7.0	8.2	9.4	11.7	—	—
	25	0.5	0.7	1.0	1.2	1.5	1.7	2.0	2.5	3.0	3.5	4.0	4.5	5.0	5.5	6.0	7.0	8.0	10.0	12.5	—
	35	0.4	0.7	0.9	1.1	1.3	1.5	1.7	2.2	2.6	3.1	3.5	3.9	4.4	4.8	5.2	6.1	7.0	8.7	10.9	—
	50	0.4	0.6	0.8	1.0	1.2	1.4	1.6	2.0	2.4	2.8	3.1	3.5	3.9	4.3	4.5	5.4	6.2	7.7	9.7	11.6
	80	0.4	0.5	0.7	0.9	1.1	1.2	1.4	1.8	2.1	2.5	2.9	3.3	3.6	4.0	4.3	5.1	5.8	7.2	9.0	10.9
	120	0.3	0.5	0.7	0.8	1.0	1.2	1.3	1.7	2.0	2.3	2.7	3.0	3.4	3.7	4.0	4.7	5.4	6.7	8.4	10.1
20	20	0.5	0.7	1.0	1.2	1.5	1.7	2.0	2.5	3.0	3.5	4.0	4.5	5.0	5.5	6.0	7.0	8.0	10.0	12.5	—
	30	0.4	0.6	0.8	1.0	1.2	1.5	1.7	2.1	2.5	2.9	3.3	3.7	4.1	4.5	4.9	5.8	6.6	8.2	10.3	12.4
	45	0.4	0.5	0.7	0.9	1.1	1.3	1.4	1.8	2.2	2.5	2.9	3.3	3.6	4.0	4.3	5.1	5.8	7.2	9.1	10.9
	60	0.3	0.5	0.7	0.8	1.0	1.2	1.3	1.7	2.0	2.3	2.7	3.0	3.4	3.7	4.0	4.7	5.4	6.7	8.4	10.1
	90	0.3	0.5	0.6	0.8	0.9	1.1	1.2	1.5	1.8	2.1	2.4	2.7	3.0	3.3	3.6	4.2	4.8	6.0	7.5	9.0
	150	0.3	0.4	0.6	0.7	0.8	1.0	1.1	1.4	1.7	2.0	2.3	2.6	2.9	3.2	3.4	4.0	4.6	5.7	7.2	8.6
24	24	0.4	0.6	0.8	1.0	1.2	1.5	1.7	2.1	2.5	2.9	3.3	3.7	4.1	4.5	5.0	5.8	6.7	8.2	10.3	12.4
	32	0.4	0.5	0.7	0.9	1.1	1.3	1.5	1.8	2.2	2.6	2.9	3.3	3.6	4.0	4.3	5.1	5.8	7.2	9.0	11.0
	50	0.3	0.5	0.6	0.8	0.9	1.1	1.2	1.5	1.8	2.2	2.5	2.8	3.1	3.4	3.7	4.4	5.0	6.2	7.8	9.4
	70	0.3	0.4	0.6	0.7	0.8	1.0	1.1	1.4	1.7	2.0	2.2	2.5	2.8	3.0	3.3	3.8	4.4	5.5	6.9	8.2
	100	0.3	0.4	0.5	0.6	0.8	0.9	1.0	1.3	1.6	1.8	2.1	2.4	2.6	2.9	3.1	3.7	4.2	5.2	6.5	7.9
	160	0.2	0.4	0.5	0.6	0.7	0.8	1.0	1.2	1.4	1.7	1.9	2.1	2.4	2.6	2.8	3.3	3.8	4.7	5.9	7.1
30	30	0.3	0.5	0.7	0.8	1.0	1.2	1.3	1.7	2.0	2.3	2.7	3.0	3.3	3.7	4.0	4.7	5.4	6.7	8.4	10.0
	45	0.3	0.4	0.6	0.7	0.8	1.0	1.1	1.4	1.7	1.9	2.2	2.5	2.7	3.0	3.3	3.8	4.4	5.5	6.9	8.2
	60	0.3	0.4	0.5	0.6	0.7	0.9	1.0	1.2	1.5	1.7	2.0	2.2	2.5	2.7	3.0	3.5	4.0	5.0	6.2	7.4
	90	0.2	0.3	0.4	0.6	0.7	0.8	0.9	1.1	1.3	1.6	1.8	2.0	2.2	2.5	2.7	3.1	3.6	4.5	5.6	6.7
	150	0.2	0.3	0.4	0.5	0.6	0.7	0.8	1.0	1.2	1.4	1.6	1.8	2.0	2.2	2.4	2.8	3.2	4.0	5.0	5.9
	200	0.2	0.3	0.4	0.5	0.6	0.7	0.8	1.0	1.1	1.3	1.5	1.7	1.9	2.0	2.2	2.6	3.0	3.7	4.7	5.6
36	36	0.3	0.4	0.6	0.7	0.8	1.0	1.1	1.4	1.7	1.9	2.2	2.5	2.8	3.0	3.3	3.9	4.4	5.5	6.9	8.3
	50	0.2	0.4	0.5	0.6	0.7	0.8	1.0	1.2	1.4	1.7	1.9	2.1	2.5	2.6	2.9	3.3	3.8	4.8	5.9	7.2
	75	0.2	0.3	0.4	0.5	0.6	0.7	0.8	1.0	1.2	1.4	1.6	1.8	2.0	2.3	2.5	2.9	3.3	4.1	5.1	6.1
	100	0.2	0.3	0.4	0.5	0.6	0.7	0.8	0.9	1.1	1.3	1.5	1.7	1.9	2.1	2.3	2.6	3.0	3.8	4.7	5.7
	150	0.2	0.3	0.3	0.4	0.5	0.6	0.7	0.9	1.0	1.2	1.4	1.6	1.7	1.9	2.1	2.4	2.8	3.5	4.3	5.2
	200	0.2	0.2	0.3	0.4	0.5	0.6	0.7	0.8	1.0	1.1	1.3	1.5	1.6	1.8	2.0	2.3	2.6	3.3	4.1	4.9
42	42	0.2	0.4	0.5	0.6	0.7	0.8	1.0	1.2	1.4	1.6	1.9	2.1	2.4	2.6	2.8	3.3	3.8	4.7	5.9	7.1
	60	0.2	0.3	0.4	0.5	0.6	0.7	0.8	1.0	1.2	1.4	1.6	1.8	2.0	2.2	2.4	2.8	3.2	4.0	5.0	6.0
	90	0.2	0.3	0.3	0.4	0.5	0.6	0.7	0.9	1.0	1.2	1.4	1.6	1.7	1.9	2.1	2.4	2.8	3.5	4.4	5.2
	140	0.2	0.2	0.3	0.4	0.5	0.5	0.6	0.8	0.9	1.1	1.2	1.4	1.5	1.7	1.9	2.2	2.5	3.1	3.9	4.6
	200	0.1	0.2	0.3	0.3	0.4	0.5	0.6	0.7	0.9	1.0	1.1	1.3	1.4	1.6	1.7	2.0	2.3	2.9	3.6	4.3
	300	0.1	0.2	0.3	0.3	0.4	0.5	0.5	0.7	0.8	0.9	1.1	1.2	1.3	1.4	1.5	1.7	1.9	2.2	2.8	3.5	4.2
50	50	0.2	0.3	0.4	0.5	0.6	0.7	0.8	1.0	1.2	1.4	1.6	1.8	2.0	2.2	2.4	2.8	3.2	4.0	5.0	6.0
	70	0.2	0.3	0.3	0.4	0.5	0.6	0.7	0.9	1.0	1.2	1.4	1.5	1.7	1.9	2.0	2.4	2.7	3.4	4.3	5.1
	100	0.1	0.2	0.3	0.4	0.4	0.5	0.6	0.7	0.9	1.0	1.2	1.3	1.5	1.6	1.8	2.1	2.4	3.0	3.7	4.5
	150	0.1	0.2	0.2	0.3	0.4	0.5	0.5	0.7	0.8	0.9	1.1	1.2	1.3	1.5	1.6	1.9	2.1	2.7	3.3	4.0
	300	0.1	0.2	0.2	0.3	0.3	0.4	0.5	0.6	0.7	0.8	0.9	1.0	1.1	1.3	1.4	1.6	1.9	2.3	2.9	3.5
60	60	0.2	0.2	0.3	0.4	0.5	0.6	0.7	0.8	1.0	1.2	1.3	1.5	1.7	1.8	2.0	2.3	2.7	3.3	4.2	5.0
	100	0.1	0.2	0.3	0.3	0.4	0.5	0.5	0.7	0.8	0.9	1.1	1.2	1.3	1.5	1.6	1.9	2.1	2.7	3.3	4.0
	150	0.1	0.2	0.2	0.3	0.3	0.4	0.5	0.6	0.7	0.8	0.9	1.1	1.2	1.3	1.4	1.6	1.9	2.3	2.9	3.5
	300	0.1	0.1	0.2	0.2	0.3	0.3	0.4	0.5	0.6	0.7	0.8	0.9	1.0	1.1	1.2	1.4	1.6	2.0	2.5	3.0
75	75	0.1	0.2	0.3	0.3	0.4	0.5	0.5	0.7	0.8	0.9	1.1	1.2	1.3	1.5	1.6	1.9	2.1	2.7	3.3	4.0
	120	0.1	0.2	0.2	0.3	0.3	0.4	0.5	0.6	0.7	0.8	0.9	1.0	1.1	1.2	1.3	1.5	1.7	2.2	2.7	3.3
	200	0.1	0.1	0.2	0.2	0.3	0.3	0.4	0.5	0.5	0.6	0.7	0.8	0.9	1.0	1.1	1.3	1.5	1.8	2.3	2.7
	300	0.1	0.1	0.2	0.2	0.2	0.3	0.3	0.4	0.5	0.6	0.6	0.7	0.8	0.9	1.0	1.2	1.3	1.7	2.1	2.5
100	100	0.1	0.1	0.2	0.2	0.3	0.3	0.4	0.5	0.6	0.7	0.8	0.9	1.0	1.1	1.2	1.4	1.6	2.0	2.5	3.0
	200	0.1	0.1	0.1	0.2	0.2	0.3	0.3	0.4	0.4	0.5	0.6	0.7	0.7	0.8	0.9	1.0	1.2	1.5	1.9	2.2
	300	0.1	0.1	0.1	0.2	0.2	0.2	0.3	0.3	0.4	0.5	0.5	0.6	0.7	0.7	0.8	0.9	1.1	1.3	1.7	2.0
150	150	0.1	0.1	0.1	0.2	0.2	0.2	0.3	0.3	0.4	0.5	0.5	0.6	0.7	0.7	0.8	0.9	1.1	1.3	1.7	2.0
	300	—	0.1	0.1	0.1	0.1	0.2	0.2	0.2	0.3	0.3	0.4	0.4	0.5	0.5	0.6	0.7	0.8	1.0	1.2	1.5
200	200	—	0.1	0.1	0.1	0.1	0.2	0.2	0.2	0.3	0.3	0.4	0.4	0.5	0.5	0.6	0.7	0.8	1.0	1.2	1.5
	300	—	0.1	0.1	0.1	0.1	0.1	0.2	0.2	0.2	0.3	0.3	0.3	0.4	0.4	0.5	0.6	0.7	0.8	1.0	1.2
300	300	—	—	0.1	0.1	0.1	0.1	0.1	0.1	0.2	0.2	0.2	0.3	0.3	0.3	0.4	0.4	0.5	0.6	0.7	0.8
500	500	—	—	—	—	0.1	0.1	0.1	0.1	0.1	0.1	0.2	0.2	0.2	0.2	0.2	0.3	0.3	0.4	0.5	0.6

Table 9-2. Percent Effective Ceiling or Floor Cavity Reflectance

% CEILING OR FLOOR REFLECTANCE	90				80				70			50			30				10		
% WALL REFLECTANCE	90	70	50	30	80	70	50	30	70	50	30	70	50	30	65	50	30	10	50	30	10
0	90	90	90	90	80	80	80	80	70	70	70	50	50	50	30	30	30	30	10	10	10
0.1	90	89	88	87	79	79	78	78	69	69	68	59	49	48	30	30	29	29	10	10	10
0.2	89	88	86	85	79	78	77	76	68	67	66	49	48	47	30	29	29	28	10	10	9
0.3	89	87	85	83	78	77	75	74	68	66	64	49	47	46	30	29	28	27	10	10	9
0.4	88	86	83	81	78	76	74	72	67	65	63	48	46	45	30	29	27	26	11	10	9
0.5	88	85	81	78	77	75	73	70	66	64	61	48	46	44	29	28	27	25	11	10	9
0.6	88	84	80	76	77	75	71	68	65	62	59	47	45	43	29	28	26	25	11	10	9
0.7	88	83	78	74	76	74	70	66	65	61	58	47	44	42	29	28	26	24	11	10	8
0.8	87	82	77	73	75	73	69	65	64	60	56	47	43	41	29	27	25	23	11	10	8
0.9	87	81	76	71	75	72	68	63	63	59	55	46	43	40	29	27	25	22	11	9	8
1.0	86	80	74	69	74	71	66	61	63	58	53	46	42	39	29	27	24	22	11	9	8
1.1	86	79	73	67	74	71	65	60	62	57	52	46	41	38	29	26	24	21	11	9	8
1.2	86	78	72	65	73	70	64	58	61	56	50	45	41	37	29	26	23	20	12	9	7
1.3	85	78	70	64	73	69	63	57	61	55	49	45	40	36	29	26	23	20	12	9	7
1.4	85	77	69	62	72	68	62	55	60	54	48	45	40	35	28	26	22	19	12	9	7
1.5	85	76	68	61	72	68	61	54	59	53	47	44	39	34	28	25	22	18	12	9	7
1.6	85	75	66	59	71	67	60	53	59	52	45	44	39	33	28	25	21	18	12	9	7
1.7	84	74	65	58	71	66	59	52	58	51	44	44	38	32	28	25	21	17	12	9	7
1.8	84	73	64	56	70	65	58	50	57	50	43	43	37	32	28	25	21	17	12	9	6
1.9	84	73	63	55	70	65	57	49	57	49	42	43	37	31	28	25	20	16	12	9	6
2.0	83	72	62	53	69	64	56	48	56	48	41	43	37	30	28	24	20	16	12	9	6
2.1	83	71	61	52	69	63	55	47	56	47	40	43	36	29	28	24	20	16	13	9	6
2.2	83	70	60	51	68	63	54	45	55	46	39	42	36	29	28	24	19	15	13	9	6
2.3	83	69	59	50	68	62	53	44	54	46	38	42	35	28	28	24	19	15	13	9	6
2.4	82	68	58	48	67	61	52	43	54	45	37	42	35	27	28	24	19	14	13	9	6
2.5	82	68	57	47	67	61	51	42	53	44	36	41	34	27	27	23	18	14	13	9	6
2.6	82	67	56	46	66	60	50	41	53	43	35	41	34	26	27	23	18	13	13	9	5
2.7	82	66	55	45	66	60	49	40	52	43	34	41	33	26	27	23	18	13	13	9	5
2.8	81	66	54	44	66	59	48	39	52	42	33	41	33	25	27	23	18	13	13	9	5
2.9	81	65	53	43	65	58	48	38	51	41	33	40	33	25	27	23	17	12	13	9	5
3.0	81	64	52	42	65	58	47	38	51	40	32	40	32	24	27	22	17	12	13	8	5
3.1	80	64	51	41	64	57	46	37	50	40	31	40	32	24	27	22	17	12	13	8	5
3.2	80	63	50	40	64	57	45	36	50	39	30	40	31	23	27	22	16	11	13	8	5
3.3	80	62	49	39	64	56	44	35	49	39	30	39	31	23	27	22	16	11	13	8	5
3.4	80	62	48	38	63	56	44	34	49	38	29	39	31	22	27	22	16	11	13	8	5
3.5	79	61	48	37	63	55	43	33	48	38	29	39	30	22	26	22	16	11	13	8	5
3.6	79	60	47	36	62	54	42	33	48	37	28	39	30	21	26	21	15	10	13	8	5
3.7	79	60	46	35	62	54	42	32	48	37	27	38	30	21	26	21	15	10	13	8	4
3.8	79	59	45	35	62	53	41	31	47	36	27	38	29	21	26	21	15	10	13	8	4
3.9	78	59	45	34	61	53	40	30	47	36	26	38	29	20	26	21	15	10	13	8	4
4.0	78	58	44	33	61	52	40	30	46	35	26	38	29	20	26	21	15	9	13	8	4
4.1	78	57	43	32	60	52	39	29	46	35	25	37	28	20	26	21	14	9	13	8	4
4.2	78	57	43	32	60	51	39	29	46	34	25	37	28	19	26	20	14	9	13	8	4
4.3	78	56	42	31	60	51	38	28	45	34	25	37	28	19	26	20	14	9	13	8	4
4.4	77	56	41	30	59	51	38	28	45	34	24	37	27	19	26	20	14	8	13	8	4
4.5	77	55	41	30	59	50	37	27	45	33	24	37	27	19	25	20	14	8	14	8	4
4.6	77	55	40	29	59	50	37	26	44	33	24	36	27	18	25	20	14	8	14	8	4
4.7	77	54	40	29	58	49	36	26	44	33	23	36	26	18	25	20	13	8	14	8	4
4.8	76	54	39	28	58	49	36	25	44	32	23	36	26	18	25	19	13	8	14	8	4
4.9	76	53	38	28	58	49	35	25	44	32	23	36	26	18	25	19	13	7	14	8	4
5.0	76	53	38	27	57	48	35	25	43	32	22	36	26	17	25	19	13	7	14	8	4

Ceiling or Floor Cavity Ratio

Table 9-3. Coefficients of Utilization

% CCR		80%		50%	
% FCR		20%		20%	
% WR		70%	50%	50%	30%
RCR	1	.60	.58	.54	.53
	2	.56	.52	.49	.47
	3	.52	.47	.45	.42
	4	.48	.43	.31	.38
	5	.45	.39	.37	.34
	6	.42	.36	.34	.31
	7	.39	.33	.31	.28
	8	.36	.30	.29	.25
	9	.34	.27	.26	.23
	10	.31	.25	.24	.21

Table 9-4. Effective Floor-Cavity Reflectance Correction Factors

Room cavity ratio	Percent effective ceiling-cavity reflectance											
	80			70			50			10		
	Percent wall reflectance											
	50	30	10	50	30	10	50	30	10	50	30	10
1	1.08	1.08	1.07	1.07	1.06	1.06	1.05	1.04	1.04	1.01	1.01	1.01
2	1.07	1.06	1.05	1.06	1.05	1.04	1.04	1.03	1.03	1.01	1.01	1.01
3	1.05	1.04	1.03	1.05	1.04	1.03	1.03	1.03	1.02	1.01	1.01	1.01
4	1.05	1.03	1.02	1.04	1.03	1.02	1.03	1.02	1.02	1.01	1.01	1.00
5	1.04	1.03	1.02	1.03	1.02	1.02	1.02	1.02	1.01	1.01	1.01	1.00
6	1.03	1.02	1.01	1.03	1.02	1.01	1.02	1.02	1.01	1.01	1.01	1.00
7	1.03	1.02	1.01	1.03	1.02	1.01	1.02	1.01	1.01	1.01	1.01	1.00
8	1.03	1.02	1.01	1.02	1.02	1.01	1.02	1.01	1.01	1.01	1.01	1.00
9	1.02	1.01	1.01	1.02	1.01	1.01	1.02	1.01	1.01	1.01	1.01	1.00
10	1.02	1.01	1.01	1.02	1.01	1.01	1.02	1.01	1.01	1.01	1.01	1.00

If the effective floor cavity reflectance is

- **30%:** Multiply the luminaire CU by the factor shown in this table.
- **23 to 29%:** Multiply the luminaire CU by the result obtained from interpolating between 1.00 and the factor shown in this table.
- **18 to 22%:** Use the luminaire CU directly; do not use this table.
- **11 to 17%:** Divide the luminaire CU by the result obtained from interpolating between 1.00 and the factor shown in this table.
- **10%:** Divide the luminaire CU by the factor shown in this table.

Courtesy Illuminating Engineering Society.

Table 9-4 shows that, for an 80% ceiling, 50% wall, and 4 room cavity ratio, the correction factor is 1.05. This will be applied to the tentative coefficient of utilization of .43 obtained previously. The instructions in Table 9-4 require interpolation between 1 and 1.05 as follows:

1. Seventeen percent is 3/10 of the way from 20% to 10%.
2. A factor of 1 applies to 20%, and 1.05 applies to 10%.
3. Therefore, the correction factor we want is 3/10 of the way from 1 to 1.05.
4. Multiplying 3/10 by (1.05 −1) gives .3 × .05 =.015.
5. Adding .015 and 1 gives 1.015 as the required correction factor.

The instructions in Table 9-4 state that the lighting fixture coefficient of utilization is to be divided by this factor. Thus, 0.43 ÷ 1.015 = 0.424, and the corrected coefficient of utilization for a 17% effective floor-cavity reflectance will be 0.42 instead of 0.43.

MAINTENANCE FACTOR

The maintenance factor takes into account the reduction in light output because of lamp aging and because of dirt accumulation. The appropriate maintenance factor for any given condition and lighting fixture type may be determined as follows.

Types of lighting fixtures are divided into five categories. The category for each lighting fixture in Appendix B is indicated in the fixture column.

After determining the category, the maintenance factor can be read from one of the five curves for each category in Fig. 9-4. The point on the curve should be selected on the basis of the estimated number of months between cleaning of the lighting fixtures. The particular curve selected should be based on the dirt content of the atmosphere under consideration.

NUMBER OF LAMPS AND LIGHTING FIXTURES REQUIRED

The number of lighting fixtures and lamps can be calculated from the following formula:

$$NF = \frac{FA \times DF}{LPF \times LPL \times CU \times MF}$$

where,
NF is number of fixtures,
FA is floor area,
DF is desired footcandles,
LPF is lamps per fixture,
LPL is lumens per lamp,
CU is coefficient of utilization,
MF is maintenance factor.

As an example, using the room as illustrated in Fig. 9-1, assume the seeing task to be 150 footcandles. We have previously determined the floor area to be 96 square feet and the coefficient of utilization to be 0.42. The lighting fixtures that will be used will contain four 40-watt fluorescent lamps. We will assume a maintenance factor of 0.7. Referring to lamp data tables, we find that

LUMEN METHOD-ZONAL-CAVITY SYSTEM

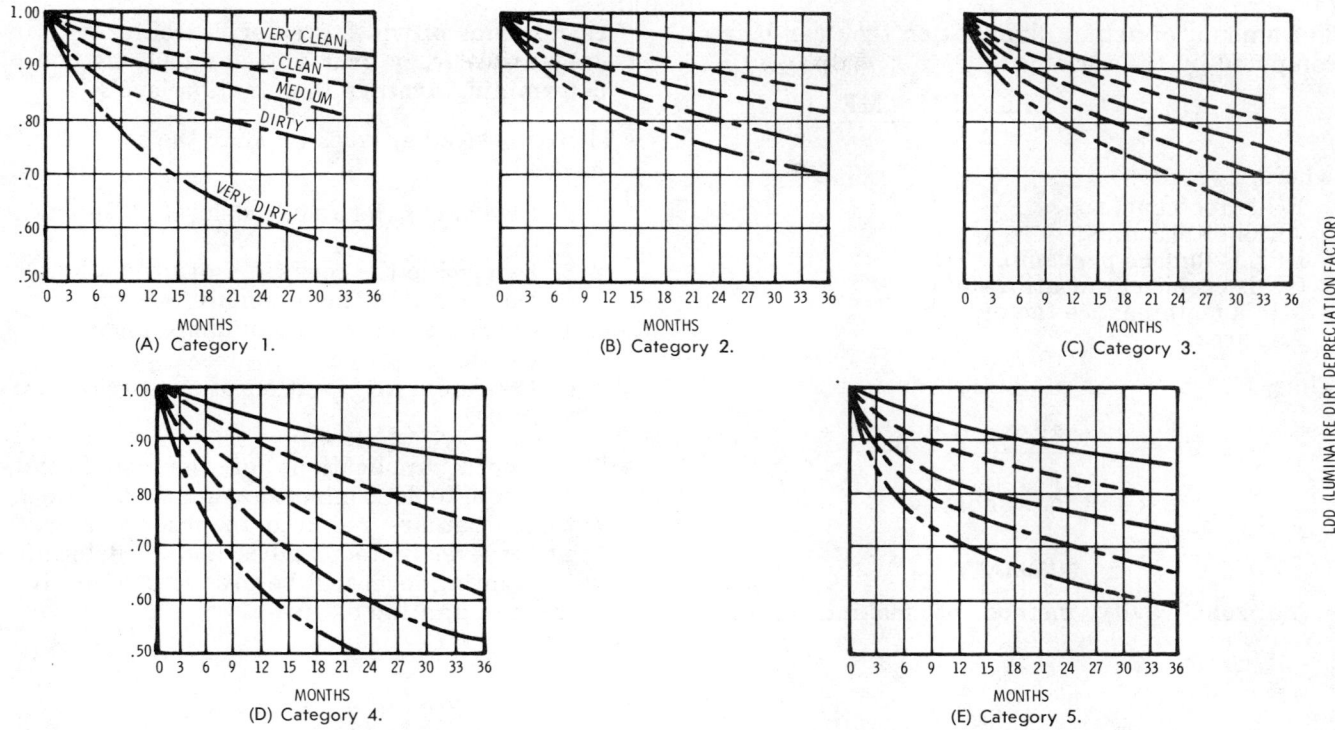

Fig. 9-4. Maintenance factor curves of the five lighting fixture categories.

this lamp has approximately 3250 initial lumens. By inserting these values in the formula, we have:

$$\text{No. fixtures} = \frac{96 \times 150}{4 \times 3250 \times 0.42 \times 0.7}$$
$$= 3.8 \text{ or } 4 \text{ fixtures}$$

We will then need four lighting fixtures to obtain approximately 150 footcandles of illumination. Since the calculation did not come out even (3.8), four fixtures will give slightly more than 150 footcandles, but this slight difference is insignificant for all practical purposes.

LAYOUT OF LIGHTING FIXTURES

Lighting fixture locations depend on the general architecture, size of bays, type of lighting fixture under consideration, etc.

In order to provide even distribution of illumination for an area, the permissible maximum spacing recommendations should not be exceeded. These recommendation ratios are supplied in terms of maximum spacing to mounting height. The "Spacing Not to Exceed" column of Appendix B gives maximum permissible ratios of spacing to mounting height above the work plane for the types of lighting fixture included. In most cases it is necessary to locate fixtures closer together than these maximums, in order to obtain required illumination levels.

In our example, the interior dimensions of the room in Fig. 9-1 is 12 feet × 8 feet and the floor plan is shown in Fig. 9-5.

We have decided to space four 2 feet × 4 feet lighting fixtures equally over this area, as also illustrated in Fig. 9-5.

ADJUSTED FOOTCANDLES

As mentioned previously, the exact number of lighting fixtures required for this room is 3.8. However, it would be very difficult indeed to install 0.8 of a lighting fixture, so 4 fixtures were used.

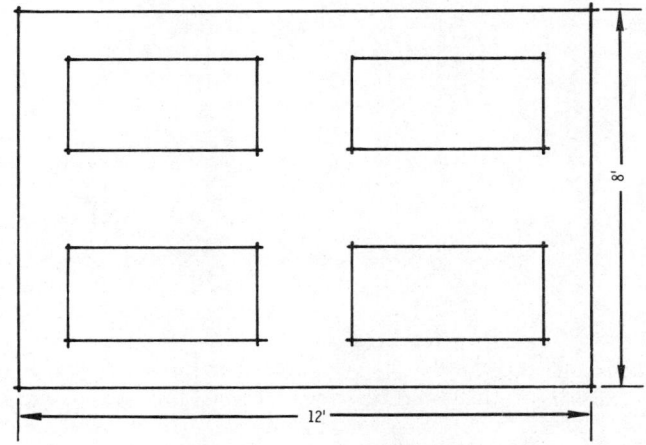

Fig. 9-5. Floor plan of room in Fig. 9-1.

The amount of actual illumination then can be recomputed by the formula:

$$FC = \frac{TL \times LPL \times CU \times MF}{A}$$

where,
FC is footcandles,
TL is total lamps,
LPL is lumens per lamp,
CU is coefficient of utilization,
MF is maintenance factor,
A is area.

Thus,

$$FC = \frac{16 \times 3250 \times 0.42 \times 0.7}{96}$$
$$= 159 \text{ footcandles}$$

SUMMARY

- The zonal-cavity method of making lighting calculations provides greater flexibility with increased accuracy over any previous method of determining average illumination levels.

- There are five key steps in using the zonal-cavity method.
  1. Determine the required level of illumination.
  2. Determine the coefficient of utilization.
  3. Determine the maintenance factor.
  4. Calculate the number of lamps and lighting fixtures required.
  5. Determine the location of the lighting fixtures.

- The general applications of the zonal-cavity method are to determine the number of lighting fixtures that are required to produce a given lighting level in footcandles and to determine what lighting level will be produced by a given number of lighting fixtures.

UNIT 10

Point-by-Point Method

Occasionally it is desirable to know what the illumination will be from one or more lighting fixtures upon a specific point or seeing task. This is especially true in designing outdoor lighting and for large warehouses or similar areas where reflectance from surrounding surfaces is insignificant.

In using the point-by-point method, a specific point is selected at which it is desired to know the illumination level, as, for example, point "P" in Fig. 10-1. Once the seeing task or point has been determined, the illumination level at the point can be calculated with the aid of a candlepower distribution curve of the lighting fixture under consideration and with the use of the following formula.

$$\text{Footcandles} = \frac{\text{Candlepower} \times \cos^3 \theta}{H^2}$$

(on horizontal plane)

It is obvious in Fig. 10-2 that the illumination at point "P," or at any other point in the area, is due to light coming from all of the lighting fixtures. In this case, the calculations must be repeated to determine the amount of light that each fixture contributes to the point; the total amount is the sum of all the contributing values.

Before attempting any actual calculations using the point-by-point method, a knowledge of candlepower distribution curves and a review of trigonometric functions is necessary.

CANDLEPOWER DISTRIBUTION CURVES

A candlepower distribution curve or graph consists of lines plotted on a polar diagram which show graphically the distribution of the light flux in some given plane around an actual light source. It also shows the apparent candlepower intensities in various directions about the light source.

Fig. 10-3 illustrates a typical candlepower distribution curve. It is read as follows.

The apparent luminous intensity directly downword is indicated by measuring off this intensity on the vertical to a given scale. Thus, XA represents, in length, the candlepower in a vertical direction directly below the light, which in this case is approximately 920 candlepower. Similarly, the distances XB, XC, XD, XE, XF, and XG represent, respectively, apparent luminous intensities in all directions around the light at angles above

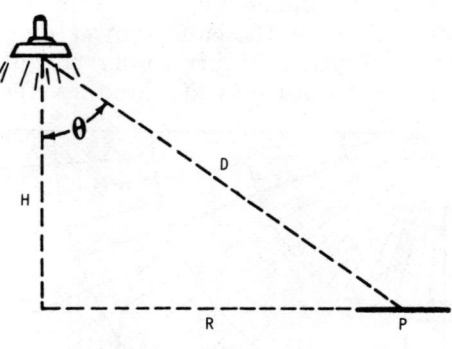

(A) Horizontal plane.

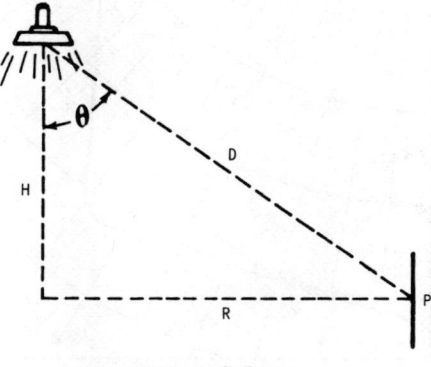

(B) Vertical plane.

Fig. 10-1. Using the point-by-point method to find the illumination level.

PRINCIPLES OF ILLUMINATION

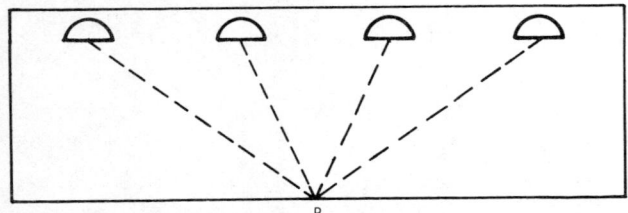

Fig. 10-2. Using the point-by-point method with more than one fixture.

the vertical of 10°, 20°, 30°, 40°, 50°, 60°, and 70°. Similarly, the apparent luminous intensities above 70° can be scaled along the lines representing their respective angles.

These points are then joined by a continuous line G, F, E, D, etc., and this line, completed for the 360°, is called the photometric distribution graph or curve of the light. The concentric circular lines (polar coordinates) are used to show the scale to which the candlepowers are plotted. The apparent candlepower intensities of the light unit can be measured along as few or as many angles as necessary, the accuracy of the resultant curve being largely determined by the number of angles taken.

The table of trigonometric functions (Table 10-1) will be helpful in determining the degrees of the angle in order to pick off the candlepower from the photometric distribution curve and also for use in the formulas.

Figure 10-3 gives the candlepower distribution curve for a lighting fixture under consideration. If this lighting fixture is mounted six feet above a seeing task or work plane, what is the horizontal illumination at a point four feet away from a point directly under the lighting fixture, as in Fig. 10-4?

First, find the angle between the perpendicular "H" and line from source to plane "D."

$$\text{tangent } \theta = \frac{R}{H} \quad \text{or} \quad \frac{4}{6} = 0.67$$

Referring to Table 10-1, 0.67 is tangent to an angle of 34°. From the same table, $\cos^3 \theta$ is 0.57. From the distribution curve in Fig. 10-3, the candlepower at 34° is approximately 920 candles. Therefore, by substituting these values into the formula given previously, we have:

$$FC = \frac{920 \times 0.57}{6 \times 6}$$

$$= 14.6 \text{ footcandles}$$

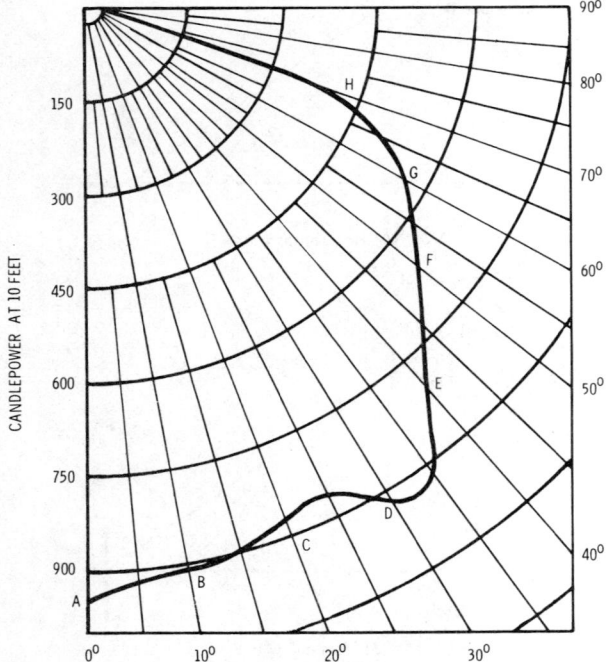

Fig. 10-3. A typical candlepower distribution curve.

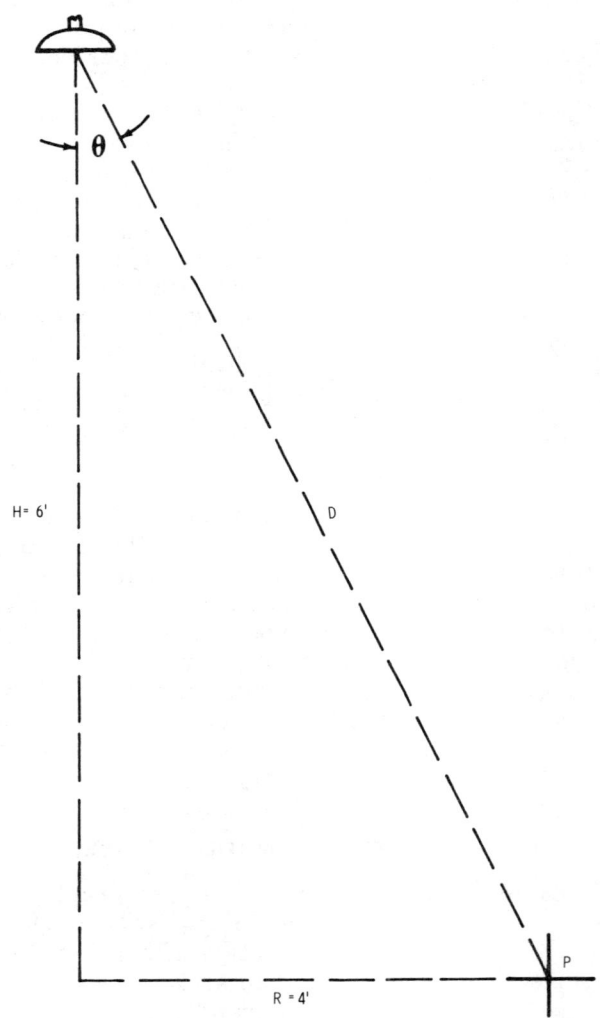

Fig. 10-4. Using the point-by-point method to find the illumination level.

POINT-BY-POINT METHOD

Table 10-1. Trigonometric Values

Angle (degrees)	Sine	Cosine	Cos²	Cos³	Tangent	Angle (degrees)	Sine	Cosine	Cos²	Cos³	Tangent
0	.0000	1.0000	1.000	1.0000	.0000	45	.7071	.7071	.500	.3536	1.0000
1	.0175	.9999	1.000	.9996	.0175	46	.7193	.6947	.483	.3352	1.0355
2	.0349	.9994	0.999	.9982	.0349	47	.7314	.6820	.465	.3172	1.0724
3	.0523	.9986	.997	.9959	.0524	48	.7431	.6691	.448	.2996	1.1106
4	.0698	.9976	.995	.9927	.0699	49	.7547	.6561	.430	.2824	1.1504
5	.0872	.9962	.992	.9886	.0875	50	.7660	.6428	.413	.2656	1.1918
6	.1045	.9945	.989	.9836	.1051	51	.7772	.6293	.396	.2492	1.2349
7	.1219	.9926	.985	.9778	.1228	52	.7880	.6157	.379	.2334	1.2799
8	.1392	.9903	.981	.9711	.1405	53	.7986	.6018	.362	.2180	1.3270
9	.1564	.9877	.976	.9635	.1584	54	.8090	.5878	.345	.2031	1.3764
10	.1737	.9848	.970	.9551	.1763	55	.8192	.5736	.329	.1887	1.4281
11	.1908	.9816	.964	.9459	.1944	56	.8290	.5592	.313	.1749	1.4826
12	.2079	.9782	.957	.9359	.2126	57	.8387	.5446	.297	.1616	1.5399
13	.2250	.9744	.949	.9251	.2309	58	.8481	.5299	.281	.1488	1.6003
14	.2419	.9703	.941	.9135	.2493	59	.8572	.5150	.265	.1366	1.6643
15	.2588	.9659	.933	.9012	.2680	60	.8660	.5000	.250	.1250	1.7321
16	.2756	.9613	.924	.8882	.2868	61	.8746	.4848	.235	.1140	1.8040
17	.2924	.9563	.915	.8745	.3057	62	.8830	.4695	.220	.1035	1.8807
18	.3090	.9511	.905	.8602	.3249	63	.8910	.4540	.206	.0937	1.9626
19	.3256	.9455	.894	.8453	.3443	64	.8988	.4384	.192	.0842	2.0503
20	.3420	.9397	.883	.8298	.3640	65	.9063	.4226	.179	.0755	2.1445
21	.3584	.9336	.872	.8137	.3839	66	.9136	.4067	.165	.0673	2.2460
22	.3746	.9272	.860	.7971	.4040	67	.9205	.3907	.153	.0597	2.3559
23	.3907	.9205	.847	.7800	.4245	68	.9272	.3746	.140	.0526	2.4751
24	.4067	.9136	.835	.7624	.4452	69	.9336	.3584	.128	.0460	2.6051
25	.4226	.9063	.821	.7444	.4663	70	.9397	.3420	.117	.0400	2.7475
26	.4384	.8988	.808	.7261	.4877	71	.9455	.3256	.106	.0345	2.9042
27	.4540	.8910	.794	.7074	.5095	72	.9511	.3090	.0955	.0295	3.0777
28	.4695	.8830	.780	.6883	.5317	73	.9563	.2924	.0855	.0250	3.2709
29	.4848	.8746	.765	.6690	.5543	74	.9613	.2756	.0762	.0209	3.4874
30	.5000	.8660	.750	.6495	.5774	75	.9659	.2588	.0670	.0173	3.7321
31	.5150	.8572	.735	.6298	.6009	76	.9703	.2419	.0585	.0142	4.0108
32	.5299	.8481	.719	.6099	.6249	77	.9744	.2250	.0506	.0114	4.3315
33	.5446	.8387	.703	.5899	.6494	78	.9782	.2079	.0432	.0090	4.7046
34	.5592	.8290	.687	.5698	.6745	79	.9816	.1908	.0364	.0069	5.1446
35	.5736	.8192	.671	.5497	.7002	80	.9848	.1737	.0302	.0052	5.6713
36	.5878	.8090	.655	.5295	.7265	81	.9877	.1564	.0245	.0038	6.3138
37	.6018	.7986	.638	.5094	.7536	82	.9903	.1392	.0194	.0027	7.1154
38	.6157	.7880	.621	.4893	.7813	83	.9926	.1219	.0149	.0018	8.1443
39	.6293	.7772	.604	.4694	.8098	84	.9945	.1045	.0109	.0011	9.5144
40	.6428	.7660	.587	.4495	.8391	85	.9962	.0872	.0076	.00066	11.430
41	.6561	.7547	.570	.4299	.8693	86	.9976	.0698	.0048	.00034	14.301
42	.6691	.7431	.552	.4104	.9004	87	.9986	.0523	.0027	.00014	19.081
43	.6820	.7314	.535	.3912	.9325	88	.9994	.0349	.0012	.00004	28.636
44	.6947	.7193	.517	.3722	.9657	89	.9999	.0175	.0003	.000005	57.290
						90	1.0000	.0000	.0000	.000000	00

If the lighting fixture in Fig. 10-4 was mounted six feet above a seeing task, what is the vertical illumination at a point four feet away from a point directly under the lighting fixture, as in Fig. 10-5? Use the candlepower distribution curve in Fig. 10-3.

From Table 10-1 the angle is approximately 34°. From the same table, cos² is 0.687 and sine is 0.559. Therefore,

$$FC = \frac{920 \times 0.687 \times 0.559}{6 \times 6}$$
$$= 9.8 \text{ or } 10 \text{ footcandles}$$

SECTION IV

INTERIOR LIGHTING DESIGN

UNIT 11

Offices and Schools

The requirements for office and school lighting are quite similar in that both require a relatively high level of illumination with good visual comfort in order to satisfy the needs of a wide range of seeing tasks over long periods of time.

The benefits of good lighting are:

1. It stimulates morale and efficiency.
2. It increases production.
3. It improves accuracy.
4. It helps conserve energy.

QUALITY OF ILLUMINATION

If an average school classroom was to be lighted with a single 1500-watt incandescent lamp hung in the center of the room, it would certainly produce a sufficient amount of illumination, but the resulting lighting effect would be intolerable. First, the intensely bright filament would repel even a glance. Second, the image of this bright light would be reflected from polished desk tops and other shiny objects. Third, persons sitting at desks and facing away from the center of the room would cast dark shadows on the desks. Fourth, the illumination level in the room would be uneven. The level directly under the lamp would be extremely high, while the light around the perimeter would be relatively weak.

The example given above illustrates what a lighting designer should strive to overcome. Office and school lighting requires that careful considertion be given to the quality of illumination which includes brightness ratios, glare, shadows, and uniformity. Color and color combinations should also be considered for best effects.

BRIGHTNESS RATIOS

Comfortable seeing conditions in offices and classrooms can be obtained only if the brightness of the light source is kept within agreeable limits. The degree of brightness control required is dependent upon the source used, the size of the room, the illumination level, the reflectances, the finish of room surfaces and furniture, and the nature of the seeing task.

A general guide to acceptable brightness limits for fluorescent fixtures in school or office areas is given in Table 11-1.

If the average brightness of a lighting fixture at each of the angles in Table 11-1 does not exceed the luminances in any single column, the brightness will meet the generally accepted limits for control of direct glare.

A comfortable balance of perceived brightness in the office requires that the brightness ratios between areas of appreciable size from normal viewpoints be with the following:

1 to 1/3—Between task and adjacent surroundings.
1 to 1/10—Between task and more remote darker surfaces.
1 to 10—Between task and more remote lighter surfaces.
20 to 1—Between fixtures and surroundings adjacent to them.

Table 11-1. Average Brightness for Various Angles

Angle	Average Brightness (Footlamberts)								
85°	250	240	230	220	210	200	190	180	165
75°	250	250	250	250	250	250	250	250	250
65°	250	265	280	295	310	325	340	355	375
55°	250	285	315	350	385	415	450	480	535
45°	250	310	365	420	480	540	600	660	750

40 to 1—Anywhere within the normal field of view.

In school rooms the acceptable ratios are:

1 to 1/3—Between task and adjacent surroundings and remote darker surfaces.
1 to 10—Between task and remote lighter surfaces at 30 footcandles. This ratio should decrease as the level of illumination increases. At 150 footcandles the ratio is 1 to 3.

RECOMMENDED REFLECTANCES

For good brightness ratios in offices and schools, the reflectances in Table 10-2 are recommended.

LIGHTING FIXTURES

Whether intended for pendant, surface, or recessed mounting, lighting fixtures should be of low brightness. For this reason, suspended fixtures providing upward light are popular. Surface-mounted fixtures may be provided with luminous side panels to reduce the brightness ratio between fixture and ceiling. Recessed fixtures should have diffusing panels that will give approximately the same brightness as that of the ceiling against which the fixture is viewed.

GLARE

Since the eye cannot render clear vision when a bright light source is within field of view, glare is one of the main factors of poor lighting. Glare reduces the sensitivity of the visual sense, and therefore reduces the visibility of a seeing task. In fact, glare is distracting and annoying, often to the extent of causing extreme discomfort and even pain.

Elimination of glare is a matter of proper brightness control. Direct glare is prevented by enclosing the loop with a diffuser or by proper shielding.

LIGHTING DESIGN

Modern lighting for offices and schools is related to the architectural design and requires careful coordination of the work of the architect and lighting designer in order to obtain a complete and satisfactory lighting system.

Table 11-2. Recommended Reflectance in Offices and Schools

Surface	Reflectance in Percent	
	Office	School
Ceiling	80-92	70-90
Walls	40-60	40-60
Desk Tops and Other Furniture	26-44	35-50
Floors	21-39	30-50

Most lighting systems for offices and schools are designed on the basis of being economically justifiable to provide a sufficient level of illumination at the proper environmental brightness. However, aesthetics should also be considered early in the planning stage.

The first point to determine is the proper amount of illumination. Appendix A gives the recommended illumination levels for public building interiors, commercial interiors, residential interiors, industrial interiors and various school areas.

The second point of design is to lay out the system; that is, to choose the lighting equipment and determine its location.

With the illumination level determined and the layout made, the next problem is to determine the number of lamps required. Unit 9 gives the procedure for calculating the number of lighting fixtures and lamps.

The next three pages show examples of the General Electric Q250 PAR-38 lamp used in school applications.

PRINCIPLES OF ILLUMINATION

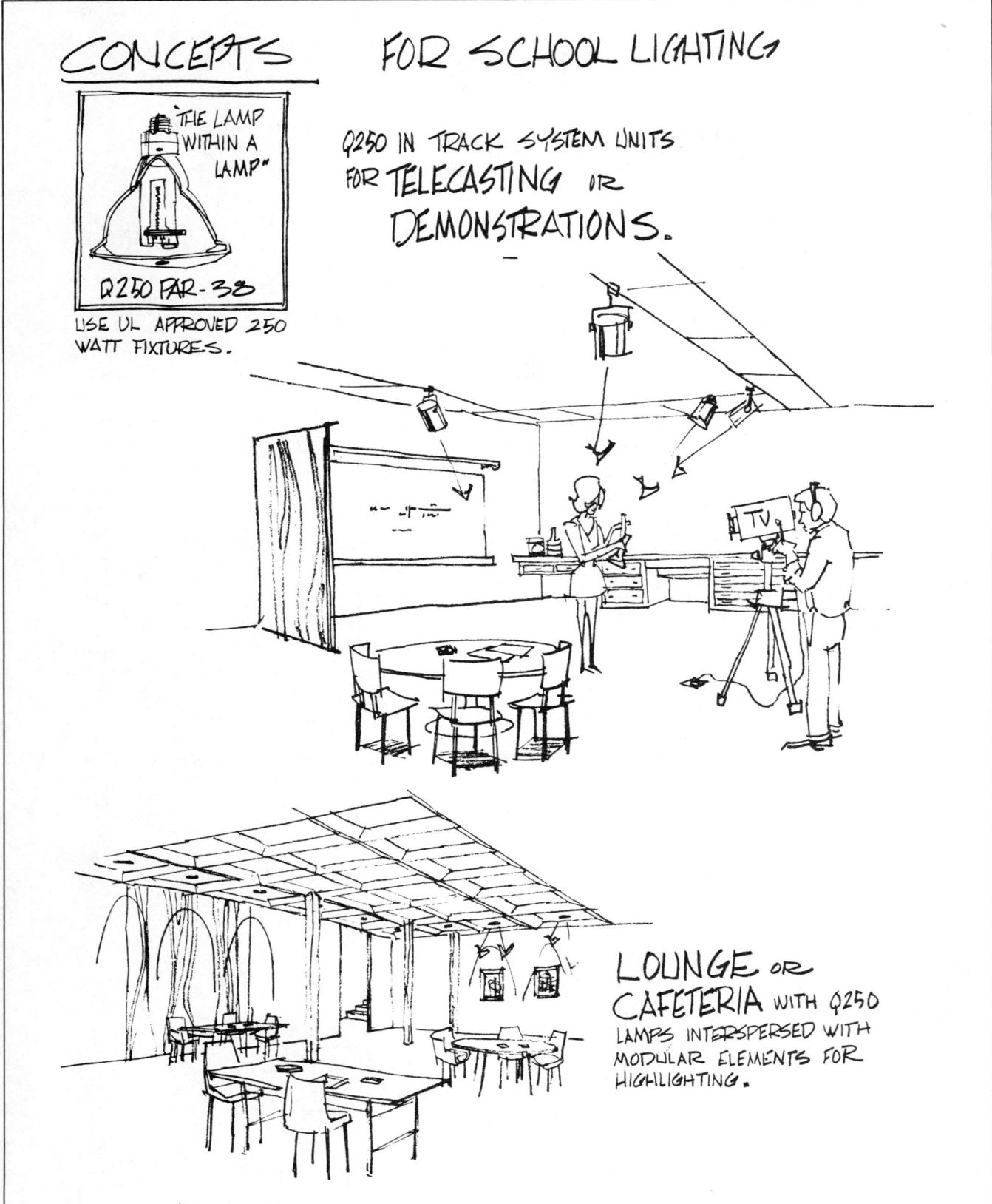

OFFICES AND SCHOOLS

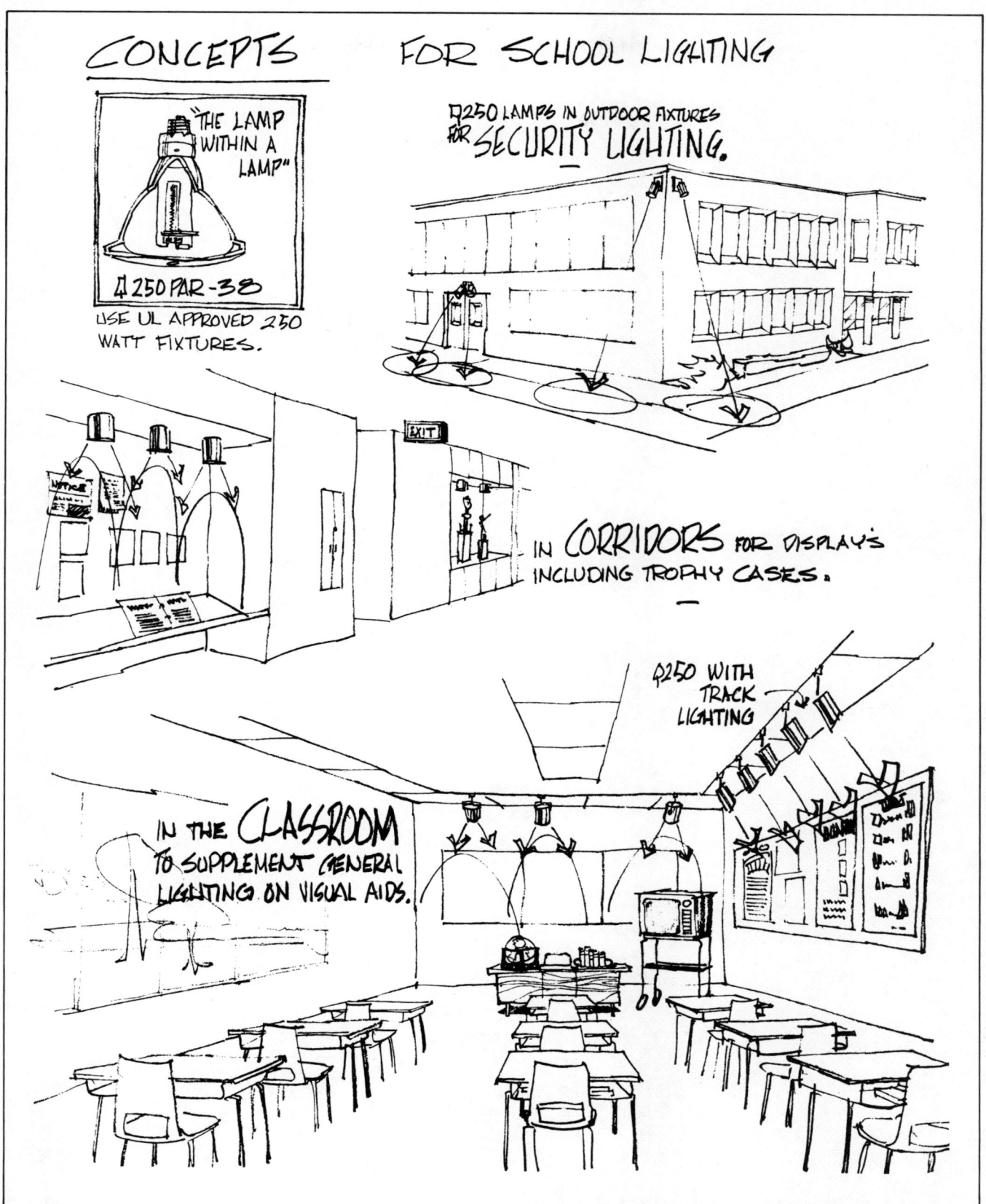

UNIT 12

Store Lighting

Light, as any store owner knows, is an important factor in selling. It is needed to attract customers by providing them with pleasant surroundings. By using the many kinds of visual impressions that can be created by lighting effects the owner may achieve a dramatic presentation of his merchandise. Light also plays an important role in improving the morale and selling efficiency of the employees.

Thus, the lighting of a store interior should combine three major functions:

1. *Attraction*—To take advantage of involuntary attraction, which is the tendency of the eye to turn instinctively toward bright areas.
2. *Appraisal*—To give the customer the proper illumination to make a full appraisal of what is seen when examining the color, texture, and quality of the merchandise.
3. *Atmosphere*—To create atmosphere or feeling in a room that appeals to the customer's emotions and to his sense of beauty.

The principles of store lighting can be applied to lighting systems for restaurants, hotels/motels, banks, and similar establishments.

ILLUMINATION LEVELS

Appendix A shows some recommended illumination levels representing modern practice in various store applications. These levels will, however, vary with the atmosphere desired. Too many lighting designs are begun with a selected lighting level which is then computed mathematically to determine the number of fixtures required. A good store lighting design should first begin by using a lighting approach that emphasizes all the basic lighting effects on merchandise, including color and brightness control. The question of directional and diffused lighting for general illumination is the second decision.

CHOICE OF LIGHT SOURCES

Some stores use fluorescent lamps almost exclusively, while others may use only incandescent lamps. High intensity discharge lighting systems are gradually finding their way into mass merchandising areas and are replacing both fluorescent and incandescent lamps. Most designers of store lighting, however, use combinations of incandescent, fluorescent, and high-intensity discharge equipment to achieve a particular objective. Basically, this objective should be to provide a lighting system that will, when tied to an overall environmental design, create an overall psychological or emotional response in the customer.

USE OF FLUORESCENT LAMPS

The wide variety of fluorescent lamps available have made this type of lamp the *favorite* for store lighting designers. These elongated light sources have made possible many applications of valance and vertical-surface lighting, cove lighting, and general illumination.

Fluorescent lamps are provided in several different qualities of *white* light, and the proper use of each is fully described in Unit 7. In general, the designations *warm* and *cool* represent the difference between artificial and natural daylight in the feeling they give to a room. Their deluxe counterparts have a greater content of red light, which is supplied by a second phosphor within the tube. While the additional red light provides a quality fluorescent light which closely approaches that of incandescent light, there is a sacrifice in efficiency.

Many colored fluorescent lamps are also available and are interchangeable with conventional white lamps. However, the relatively large size and low brightness of fluorescent lamps limit the total amount of colored light that can be delivered from a given source, and accurate control of the direction of the light is difficult. Therefore, colored fluorescent lamps are more useful for flooding large areas than for small areas requiring concentrated light.

PROJECTOR AND REFLECTOR LAMPS

Projector and reflector incandescent lamps have proved useful for many special applications in the lighting of stores. All such lamps used in store areas should be carefully aimed to avoid creating annoying glare or reflections from mirror-like surfaces. This type lighting, when properly used, adds greatly to the sales appeal, decor, and visual environment of store areas.

Reflector color lamps are a convenient and effective source of colored light, especially where control of such light is needed. There are at least six colors available, all of which are designed for maximum effectiveness when used either alone or in combination with each other. For example: pink and blue-white lamps can be used to give an entire display a warm or cool tone with very little color distortion; pink lamps complement complexion tones; blue-white lamps are excellent for lighting displays of silverware, jewelry, or home appliances.

The sketches in Figs. 12-1 through 12-4, furnished by General Electric, illustrate some of the uses of their new Q250 Par-38 quartzline lamps in store areas, office lobbies and reception areas, banks, and art museums and galleries.

COLOR CONSIDERATION

Color probably presents the biggest design problem in lighting store areas, although the size, shape, and finish of merchandise will also influence the choice of light sources and lighting fixtures in a given store area. For example, a deep-pile rug has a textured finish, and evenly diffused light may illuminate each fiber so uniformly that appraisal of the pattern and fiber depth may be difficult. However, directional and diffused light used together would emphasize the soft texture and deep nap of the rug.

Many kinds of merchandise show up better under a dual system of incandescent/quartz and fluorescent lighting than under one system or the other. Therefore, both types are normally used in the majority of store areas. In the overall lighting of stores, fluorescent lamps are used where the goal is a cool atmosphere or greatest economy. A predominantly incandescent system may be used where a warmer atmosphere is desired, although deluxe warm white fluorescent may be a better choice for this effect.

A guideline for choice of light sources in stores where color is a determining factor is that lighting which shows merchandise as it will appear in actual use. This method will usually bring customer satisfaction and will minimize return of goods because of color change at the point of use.

LIGHTING FOR EMPHASIS OF MERCHANDISE

Light is one of the most effective means of directing attention to merchandise. The cardinal principle in store lighting is to make the merchandise brighter than other areas in the field of view. To this end, the light should be directed to the display counters, showcases, wall cases, and feature displays, rather than to circulation areas. The brightnesses of luminaires, ceilings, columns, and other architectural features should be limited to relatively low values. Higher wall brightnesses are sometimes useful, however, for attracting customers to perimeter areas of the store.

Counter Areas and Showcase Tops

In most stores, counter display areas and showcase tops are satisfactorily lighted with 100 to 200 footcandles, depending on the productivity expected, the general lighting level, and the size of the details that must be readily visible. The display area illumination should be three to ten times the level of the general area lighting. Most stores display high-profit or impulse items at the end of the counter or near the cash register. These preferred locations should have two to five times as much illumination as the rest of the counter. Incandescent spotlights with their accompanying highlights and shadows are very effective.

Showcases

A showcase interior should have more illumination than the top of the case, but not more than about twice as much. Where the illumination on top is in the 30- to 75-footcandle range, T-6 or T-8 slimlines, operated at 200 milliamperes, provide sufficient light inside the case; for higher footcandles on the top, an operating current of 300 milliamperes is sometimes recommended. Too great a differential between the illumination inside the case and that on the top, where the merchan-

STORE LIGHTING

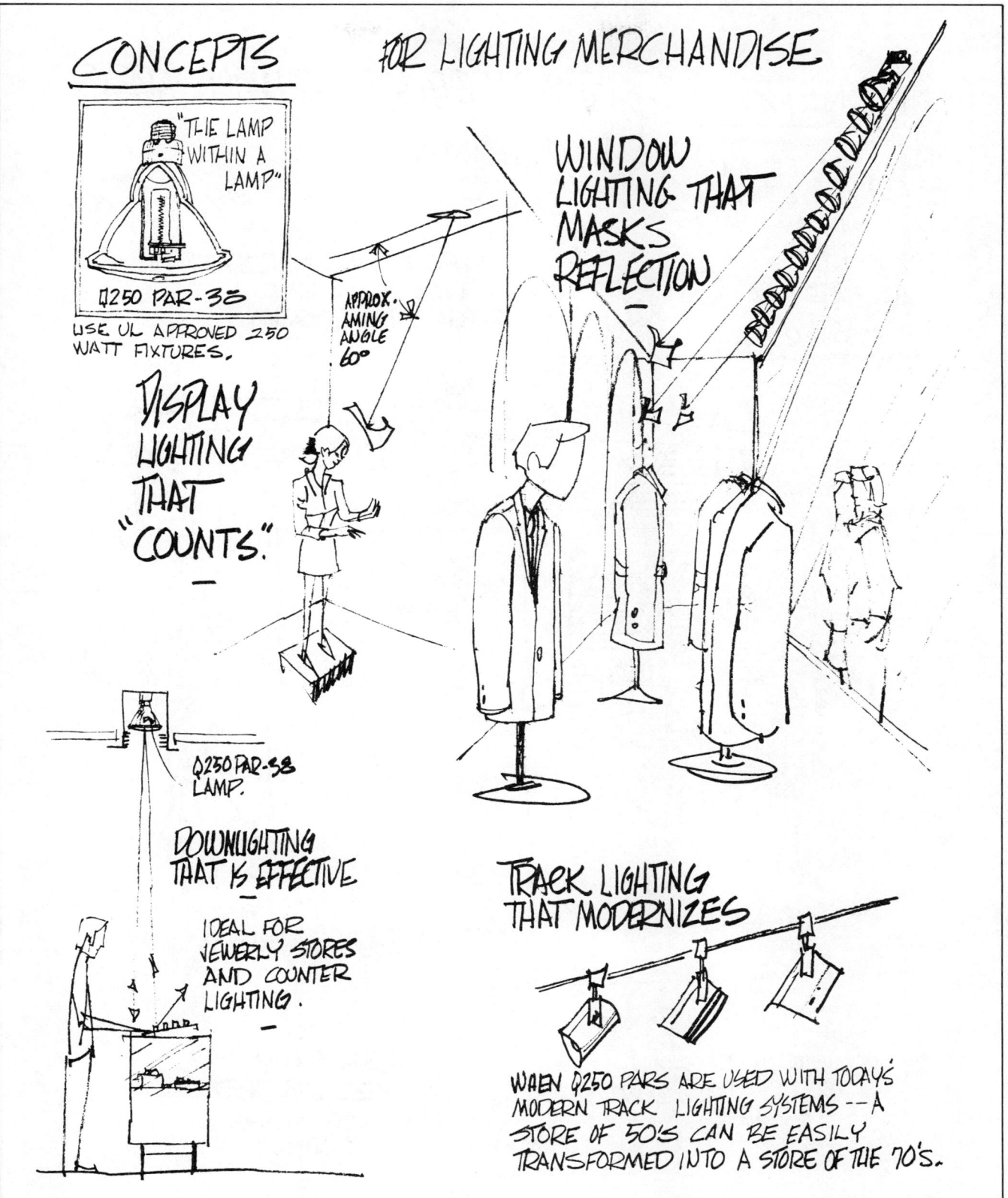

Fig. 12-1. Some ways the General Electric Q250 Par-38 quartzline lamps can be used in store areas.

PRINCIPLES OF ILLUMINATION

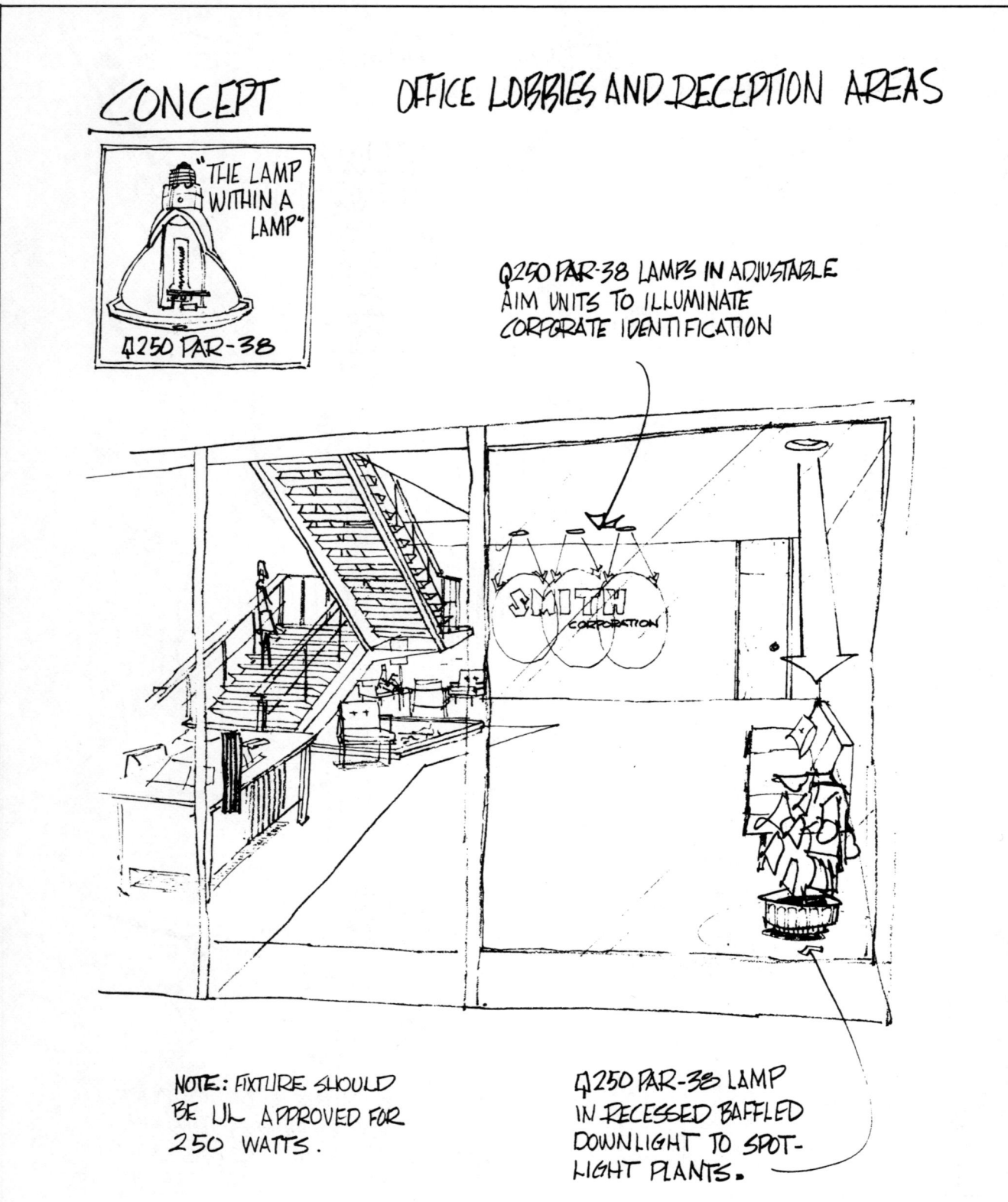

Fig. 12-2. The General Electric Q250 Par-38 quartzline lamps used in office lobbies and reception areas.

STORE LIGHTING

Fig. 12-3. High ceiling applications of the General Electric Q250 Par-38 quartzline lamps in bank with recessed downlights.

PRINCIPLES OF ILLUMINATION

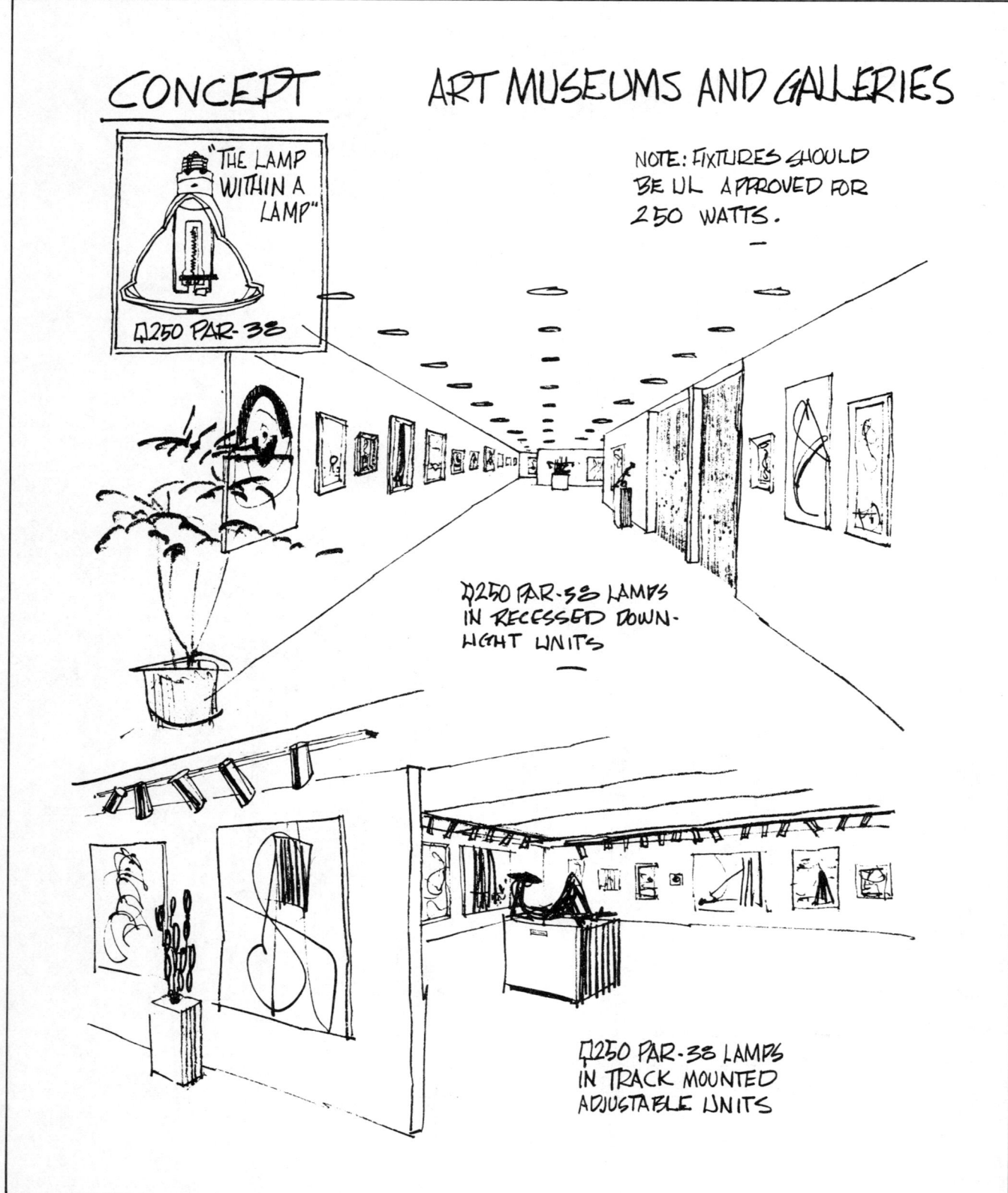

Fig. 12-4. The General Electric Q250 Par-38 quartzline lamps used in art museums and galleries.

dise is normally inspected more closely prior to purchase, is not advisable. Many products examined under an illumination level significantly lower than that under which they were displayed lose some of their attraction. This is not a consideration with feature displays, since they are seldom removed for inspection.

Wall Cases

Wall cases generally should be illuminated to approximately the same level as counter display areas. However, higher levels are desirable where the brightness of the wall case and the adjoining sales space is to be used partly to pull traffic into the area. One of the most important features of the lighting of wall cases for wearing apparel is that it be designed to light the entire case from top to bottom, rather than just the shoulders of the garments. The illumination at the bottom should be at least 1/10 of that at the top.

Table 12-1. Approximate Illumination (in Footcandles) of Fluorescent Lamps

Vertical Distance	Horizontal Distance							
	Symmetrical Reflector				Without Reflector			
	6"	12"	18"	24"	6"	12"	18"	24"
6"	21.4	15.7	12.1	9.8	30.6	22.8	16.5	13.5
12"	12.0	12.8	12.1	10.2	11.3	14.7	13.6	12.2
18"	7.4	9.6	10.0	9.1	5.2	8.8	10.2	9.6
24"	4.4	6.9	8.8	7.7	2.6	5.0	6.6	7.2
30"	2.6	4.8	6.9	6.8	1.4	3.3	4.4	5.2
36"	2.0	3.8	5.2	6.1	1.0	2.2	3.2	4.1
42"	1.4	2.8	4.1	5.0	0.6	1.4	2.2	3.0
48"	0.8	1.9	2.8	4.1	0.4	1.0	1.6	2.2
54"	0.8	1.6	2.4	3.4	0.4	0.8	1.4	1.6
60"	0.6	1.2	1.9	2.6	0.3	0.8	1.0	1.4

Table 12-1 lists the approximate illumination in footcandles produced by fluorescent lamps on a vertical surface for each 100-rated lamp lumens per foot of case length. The specular symmetrical reflector was so adjusted that its maximum candlepower was directed at a point 42 inches below the lamp. In the experiment where no reflector was used, the entire inner surface of the valance was painted white.

Feature Displays

Feature displays add sparkle and drama to the store. They not only serve to exhibit the product on display, but they also attract shoppers into their area. Feature displays at normal eye level may be satisfactorily illuminated with two or three times as much light as other displays. To be certain to capture attention, elevated displays outside the normal field of view require three to five times as much illumination as nonfeature displays. Colored light may sometimes be used effectively.

Sales Areas

In many stores, sales areas and display areas are practically the same, and their lighting becomes doubly important. Where the two do not coincide (as, for example, the areas near garment cases, which are sales areas but not actually display areas), care must be taken to ensure complete lighting coverage.

In general, the quality of the sales area lighting should be the same as that of the display lighting. For both applications direct filament lighting is well suited for most products. The filament is a small source easily controlled either by reflectors or by lenses, thus permitting a concentration of light on the proper areas. When combined with a lower level of diffused light, directional illumination is a distinct asset for lighting many kinds of products, emphasizing the weave and texture of textiles and the form and shape of three-dimensional objects. The illumination level should be adequate for rapid and accurate inspection of the merchandise. Certain articles, such as jewelry and precious stones, are most attractive when illuminated to over 100 footcandles.

Where possible, the sales area lighting should be similar in color-rendering properties to that under which the merchandise will be used. Incandescent lighting is well suited for furniture and some wearing apparel. Outdoor sports clothing and furs are usually lighted with deluxe cool white or cool white fluorescent lamps.

Mirrors

The mirror area in a clothing store is the most important sales area—the place of final inspection. The lighting must flatter not only the merchandise, but the wearer as well. A new dress, a hat, or a suit becomes a frame or background against which the purchaser views her face. The best type of lighting is usually provided by a combination of diffused light, to eliminate harsh shadows, and directional light to emphasize the facial features and the texture of fabrics, and to avoid the flat appearance of objects lighted entirely by diffused light.

The light source should be selected to flatter the skin and to provide good rendition of all colors. Deluxe fluorescent lamps are recommended for the diffuse lighting. To further enhance the appearance of the skin, some stores employ a green rug and decorate the walls near the mirror in green. Green is the complementary color of skin tones, and the face appears pink and healthy against a green background.

PRINCIPLES OF ILLUMINATION

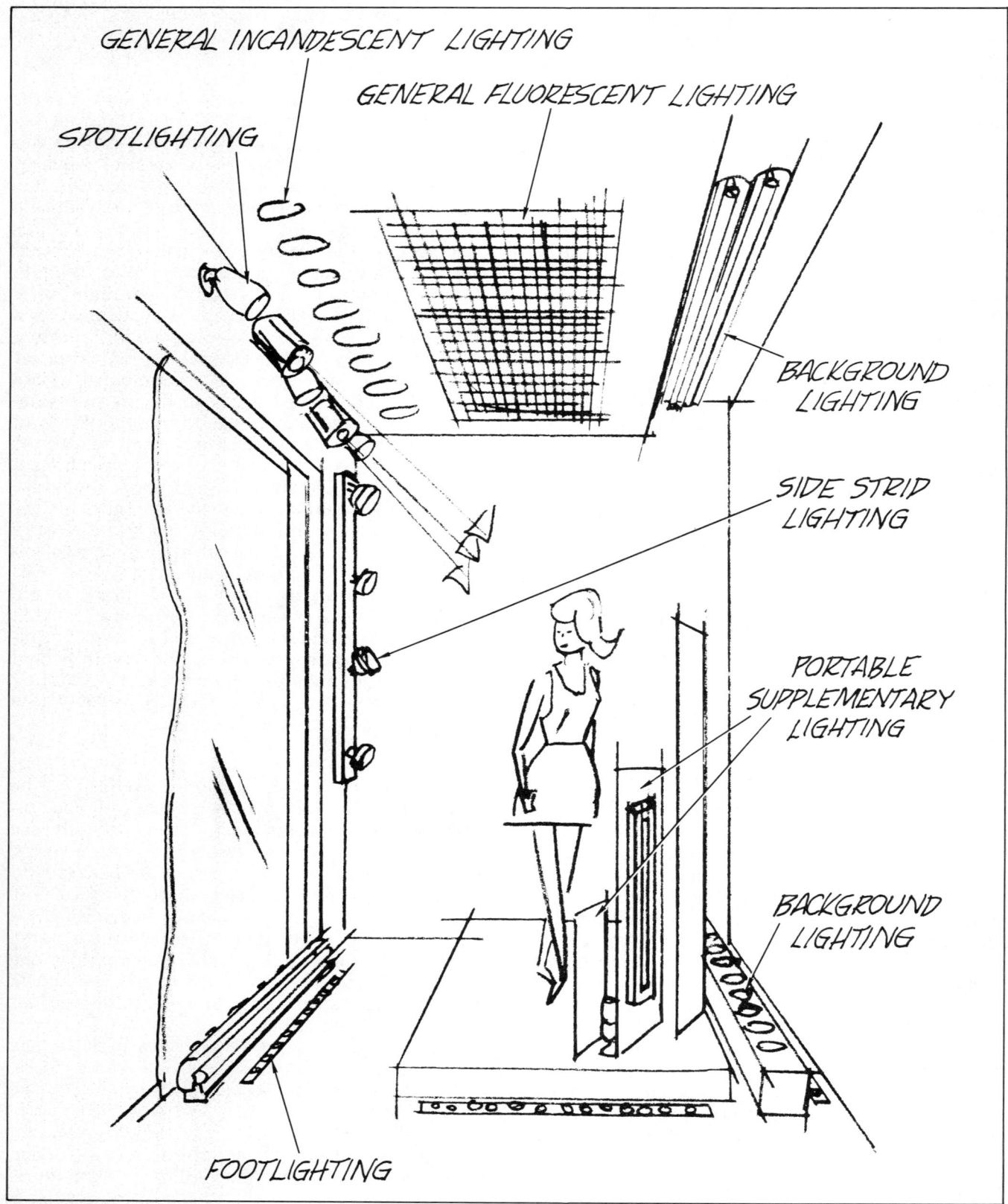

Fig. 12-5. A diagram of a balanced store lighting system.

SHOW WINDOW LIGHTING

The space occupied by the show window is the most valuable area in the store, but without good lighting its usefulness as an advertising medium can be almost lost. For the show window to function at optimum efficiency, it must capture the attention and interest of prospective shoppers. Brightness is invaluable in attracting attention, but without inviting, interesting, or dramatic displays the window may receive only a glance. On the other hand, an otherwise interesting display may go unnoticed if it is not sufficiently bright. The control of brightness, highlight, and shadow is the key to a successful window display.

Show window lighting equipment should be flexible enough to spotlight a small display vividly, to illuminate a larger area with directional light, or to flood the entire window with nondirectional diffused light, as from fluorescent luminaires (Fig. 12-6). Many window decorations require a combination of all three systems to display a single item. Dimming devices and flexible switching are valuable assets; colored lens caps for reflector and projector lamps are frequently necessary.

REFLECTED IMAGES

Large reflected images that are relatively uniform in brightness, as for instance the sky, raise the apparent brightness of those parts of the display that are in relative shadow, thus tending to destroy contrasts that may have been planned for spectacular effect. Images of isolated bright surfaces, such as the sun reflected off automobile chrome, cause unnatural highlights on the display and are more confusing and annoying than images of uniform brightness.

For these reasons it is difficult to lay down general rules for illumination levels in show windows. Each window is a special case, and must be studied in relation to its orientation and surroundings.

UNIT 13

Residential Lighting

Properly designed lighting is one of the greatest comforts and conveniences that any home owner can enjoy. In building new homes or remodeling old ones, the lighting should be considered equally as important as the heating system, the furniture placement, and as one of the most important features of both interior and exterior decorations.

As a rule, residential lighting does not require a large quantity of elaborate calculations, as does a school or office building. However, lighting designers must apply their talent and ingenuity in selecting the best types of lighting fixtures for various locations in order to obtain a desirable effect, as well as the proper amount of illumination at the desired quality. Light has certain characteristics that can be used to change the apparent shape of a room, to create a feeling of separate areas within one room, or to alter architectural line, form, color, pattern, or texture. Light also affects the mood and atmosphere within the area where used.

It should now be apparent that the various possibilities of residential lighting are limited only by the ingenuity of the lighting designer who is equipped with imagination, familiarity with residential lighting equipment, and knowledge of the basic principles of good lighting design.

In most homes today, there are 32 basic structural areas which need basic lighting prior to furniture placement. This unit covers the 32 areas, the function of each, and recommended lighting for each.

ENTRY HALL

Entry lighting covers the entire foyer area and should be carefully planned, since this is where guests receive their first impression of a house. Illumination levels vary greatly in the entry, as it is dependent upon the mood desired. However, it should be adequate for greeting guests and allowing for coat removal.

The first step is usually a centrally located ceiling-mounted fixture. This fixture may be recessed, but is generally more effective when surface-mounted or suspended.

Entry planters or other special structural areas should be accented with supplementary lighting. Recessed downlights and wallwashers are both very effective for accent lighting, as are pendant or cluster stylings. Luminous wall panels are also very effective as room dividers from the foyer to living room.

LIVING ROOM

As the living room is the social heart of most homes, lighting should emphasize special architectural features such as fireplace, bookcases, paintings, etc. The same is true of draperied walls, planters, or any other special room accents.

Dramatizing fireplaces with accent lights brings out texture of bricks, adds to overall room light level, and eliminates bright spots that cause subconscious irritation over a period of time. Use 75- to 150-watt lamps in wallwash-type fixtures—either recessed or surface mounted—for this application.

While recessed downlights, cornice, or valance lighting all add life to draperies, they also supplement the general living room lighting level. Position downlights 2½ to 3 feet apart and 8 to 10 inches from the wall. Valances are always used at windows, usually with draperies (Fig. 13-1). They provide up-light, which reflects off ceiling for general room lighting, and down-light for drapery accent. Cornices direct all their light downward to give dramatic interest to wall coverings, draperies, etc., and are good for low-ceiling rooms (Fig. 13-2).

PRINCIPLES OF ILLUMINATION

Fig. 13-1. Valences provide up-light which reflects off ceiling for general room lighting and down-light for drapery accent.

Pulldown fixtures or table lamps are used for reading areas. While the pulldown fixtures are more dramatic, the designer must know the furniture arrangement prior to fixture placement.

As a final touch, add dimmers to vary the lighting levels exactly to the living room activities—low for a relaxed mood, bright for a gay, party mood.

DINING ROOM

A chandelier above the dining table (should be centered) for general illumination becomes the *centerpiece* of the room. The dimming of the lamps in the chandelier adds versatility to the dining room. It sets the mood of the activity: low candlelight effect for formal dining, bright for an evening of cards.

Good planning also calls for supplementary lighting at buffet and sideboard areas. For a contemporary design, use recessed accent lights in these areas; use wall brackets to match chandelier for a traditional setting. Additional supplementary lighting may be used around windows in valance or cornice effects using concealed fluorescent lighting.

Fig. 13-2. Cornices direct all their light downward.

RESIDENTIAL LIGHTING

KITCHEN

The kitchen is the basic and most-used work area of any home, and carefully planned lighting should be utilized to make day-to-day tasks easier.

Good kitchen lighting begins with a central light source. Luminous ceilings provide skylight effect and are probably the best "working lights." They are also the most expensive and are not within the budget of every homeowner. Use bare fluorescent strips (preferably with dimming controls) above ceiling panels. Choose panels with attractive diffuser patterns.

If the luminous ceiling is not practical, surface-mounted decorative incandescent or fluorescent fixtures are fine. If fluorescent lamps are used, deluxe warm-white lamps should be utilized, as food and people will have a much better appearance than with standard cool-white lamps.

Kitchen work can be made much easier if there is other supplementary lighting for certain tasks. For example, a recessed downlight or soffit lighting above the sink, shallow undercabinet fixtures, and light over range (usually built into the range hood) all provide good supplementary lighting to eliminate shadows.

If the kitchen has a dinette, use a pulldown or pendant-hung fixture directly over the table for both functional and decorative illumination.

BATHROOM

Lighting performs a wide variety of specific tasks in bathrooms. More so than one might imagine.

For good grooming, the lavatory-vanity should be lighted to remove all shadows from faces. Mirror lights (wall bracket, lighted soffits, downlights) give the best results. Over a vanity table, plan pendants or downlights for concentrated light with a decorative touch. For safety and health, a moisture-proof, ceiling-mounted recessed light should be planned above bath or shower. Linen closet lighting should also be considered.

The bathroom needs as much general lighting as any other room. If the bath is small, usually the mirror and tub or shower lights will also suffice for general illumination. However, in the larger baths, a bright central source (recessed, surface and decorative fluorescent luminous ceiling) is needed to transform it from the dim bath of the past to a smart, bright part of the house today.

BEDROOM AND HALLS

Central room lighting will give an equal amount of light to all corners of the bedroom for dressing and cleaning, as well as caring for the sick. A compact, close-to-ceiling decorative glass fixture is recommended. This one fixture, however, cannot handle the entire needs of the room. Light is also needed for grooming and reading. Treat vanity mirrors as separate areas with their own lighting, to make good grooming easier.

Bedroom closets, as well as all other closets in the home, should have recessed lighting fixtures using 60- to 100-watt lamps and equipped with fresnel lenses. The closet lights can be installed with a single-pole switch outside of the closet area. It is also good to provide switches that have glowing handles when the lamp is on. This prevents leaving the closet lights burning accidentally; the glowing handle indicates that the lamp is on.

Wall lights with directional beams are good for bedtime reading, as are adjustable-height pulldowns over each bedstand.

Halls and corridors require general illumination along their entire lengths. Recessed fixtures with fresnel lenses are excellent for contemporary homes, while wall brackets may fit in more with traditional homes. Wall-hung pictures may be accented with recessed or surface-mounted spotlight fixtures.

FAMILY ROOM

Recently, the family room has become basic to home planning. Separate from more formal areas of the house, it allows plenty of space for family relaxation, games, and hobbies. Because of this versatility, the family room requires a variety of lighting effects.

General illumination should be considered first. Since many family rooms are in the basement with its low ceiling, recessed or very compact fixtures are necessary. But there is an endless number of patterns that can be used with these fixtures. A luminous ceiling is also good if controlled by dimmers. Much depends on the room layout and its uses, as well as the family's needs.

Additionally, after the general lighting has been planned, consider special lighting for activity areas as a practical decorative touch. A centered pulldown fixture turns a table into a special game area. Hobby areas should also be lighted with appropriate lighting. All light sources should be on dimmers to match lighting levels to the room activities.

OUTDOOR LIGHTING

Outdoor lighting should continue the general architectural theme of the house—contemporary, colonial, etc. This phase of lighting includes that

PRINCIPLES OF ILLUMINATION

for: front entry, walk, eaves of house, patio, back entry, porch, garage, and driveway.

Outside wall bracket fixtures around the door welcome guests, as well as post lights or low-level lights along walkway. Recessed fixtures in soffits are good to brighten the walk, while they also tend to dramatize walls.

Outdoor lighting not only welcomes guests and lights their paths, but it also creates a hospitable look and turns the yard and patio into extra living area during the summer. As a safety factor, it discourages prowlers and helps reduce accidents.

A pair of outdoor wall bracket fixtures, a ceiling pendant fixture, or a recessed fixture at front and rear doorways are basic. If the walk is long, include a decorative, yet practical, post-mounted fixture. Post lights can be used in conjunction with a time-clock or an electric-eye switch to light at dusk and go off at dawn. In fact, entire subdivisions have been lighted in this manner.

For decorative and safety reasons, install recessed or outdoor floodlights at the house corners and in or on the soffit. Outdoor patios may be lighted with wall bracket fixtures or floodlights mounted on the house, or post lights. A combination of all three offers great versatility and provides for changes in moods.

The garage is a separate lighting area. Use ceiling-mounted fixtures inside it to guide your way and wall brackets beside the door or recessed lights in the soffit above the door.

Manufacturers of residential lighting equipment offer excellent catalogs and planning guides which can be the source of tremendous assistance to the lighting designer. A visit to a residential light showroom for an on-the-spot demonstration of the various fixtures available is also invaluable.

UNIT 14

Industrial Lighting

The basic objectives of good industrial lighting design are the same as in other seeing areas: to provide adequate illumination and to make that illumination comfortable to the eyes. In order to obtain this condition, the following lighting demands must be met:

1. The work plane must be adequately illuminated.
2. Shadows falling on the work plane must be eliminated or at least minimized.
3. Objectionable glare or reflections from machines, equipment, and other nearby objects or surfaces must be avoided.
4. The contrast between the brightness of the work plane and immediate surroundings should not differ greatly. Research has demonstrated that, for best seeing, the brightness of the surrounding areas should approach that of the work.

One look at the recommended levels of illumination for industrial applications in Appendix A will illustrate the wide variety of visual tasks involved in industrial lighting. The level may be 5 footcandles for a warehouse or go over 1000 footcandles for critical inspection work. The work plane may be the horizontal, the vertical, a plane at any angle, or a combination of these planes. A wide variety of material colors and other different characteristics of materials add to the problems confronting the lighting designer. However, there are many good tools and techniques available to the designer for solving these problems; a few of these will be discussed in this unit.

Good lighting will always be a basic requirement in obtaining greater accuracy and workmanship in industries. It also promotes good housekeeping, safety, and morale.

FACTORS OF GOOD LIGHTING

In designing a proper lighting system for an industrial area, the seeing task involved must first be analyzed to determine the amount and type of illumination that will provide the best visibility. Lighting equipment can then be selected that will provide the needed illumination. The following factors affect the seeing task in an industrial area:

1. Size
2. Brightness
3. Contrast
4. Time

Review Unit 2 for further explanation of these factors. While the importance of these factors will vary with each application, they should never be neglected.

Reflected Glare

Reflected glare from shiny work surfaces interferes with seeing. Even when a bright image is not actually in the work area, it may be distracting and cause discomfort due to the fact that the eyes involuntarily turn toward a bright or conspicuous object within the field of vision. Higher illumination levels reduce the hazard of reflected glare by lessening the contrast between adjacent surfaces. Large-area, low-brightness sources also reduce reflected glare.

Shadows

Shadows or differences in the brightness of the various surfaces give form to three-dimensional objects. They play an important part in seeing. However, harsh, uncontrolled shadows are undesirable, since they distract the workman and make seeing uncertain, both of which consumes

Table 14-1. Lighting Fixture Selector

Part I—Flat Surfaces

General Characteristics	Description	Lighting Requirements	Luminaire Type*
A. OPAQUE MATERIALS			
1. Diffuse detail & background			
a. Unbroken surface	Newspaper proofreading	High visibility with comfort	B or C
b. Broken surface	Scratch on unglazed tile	To emphasize surface break	A
2. Specular detail & background			
a. Unbroken surface	Dent, warps, uneven surface	Emphasize unevenness	E
b. Broken surface	Scratch, scribe, engraving, punch marks	Create contrast of cut against specular surface	C or D, E
c. Specular coating over specular background	Inspection of finish plating over underplating	To show up uncovered spots	D
3. Combined specular & diffuse surfaces			
a. Specular detail on diffuse, light background	Shiny ink or pencil marks on dull paper	To product maximum contrast without reflected glare from shiny markings	C or D
b. Specular detail on diffuse, dark background	Punch or scribe marks on dull metal	To create bright reflection from detail	B or C
c. Diffuse detail on specular, light background	Graduations on a steel scale	To create a uniform, low brightness reflection from specular background	D or C
d. Diffuse detail on specular, dark background	Wax marks on auto body	To produce high brightness of detail against dark background	C or B
B. TRANSLUCENT MATERIALS			
1. With diffuse surface	Frosted or etched glass or plastic, lightweight fabrics, hoisery	Maximum visibility of surface detail / Maximum visibility of detail within material	
2. With specular surface	Scratch on opal glass or plastic	Maximum visibility of surface detail / Maximum visibility of detail within material	
C. TRANSPARENT MATERIALS			
Clear material with specular surface	Plate glass	To produce visibility of details within material, such as bubbles, and details on surface, such as scratches	E or A
D. TRANSPARENT OVER OPAQUE MATERIALS			
1. Transparent material over diffuse background	Instrument panel	Maximum visibility of scale and pointer without reflected glare	A
	Varnished desk top	Maximum visibility of detail on or in transparent coating or on diffuse background	
		Emphasis of uneven surface	E
2. Transparent material over a specular background	Glass mirror	Maximum visibility of detail on or in transparent material	A
		Maximum visibility of detail on specular background	E

Part II—Three Dimensional Objects

General Characteristics	Description	Lighting Requirements	Luminaire Type*
A. OPAQUE MATERIALS			
1. Diffuse detail & background	Dirt on a casting or blow holes in a casting	To emphasize detail with a poor contrast	C or B or A / C or B as a black light source when object has fluorescent coat
2. Specular detail & background			
a. Detail on the surface	Dent on silverware	To emphasize surface unevenness	E
	Inspection of finish plating over underplating	To show up areas not properly plated	D
b. Detail in the surface	Scratch on a watch case	To emphasize surface break	D

Table 14-1. Lighting Fixture Selector—cont

General Characteristics	Description	Lighting Requirements	Luminaire Type*
3. Combination specular & diffuse			
a. Specular detail on diffuse background	Scribe mark on casting	To make line glitter against dull background	C or B
b. Diffuse detail on specular background	Micrometer scale	To create luminous background against which scale markings can be seen in high contrast	D or C
	Coal picking	To make coal glitter in contrast to dull impurities	A or B
B. TRANSLUCENT MATERIALS			
1. Diffuse surface	Lamp shade	To show imperfections in material	B
2. Specular surface	Glass enclosing globe	To emphasize surface irregularities	E
		To check homogeneity	B
C. TRANSPARENT MATERIALS			
Clear material with specular surface	Bottles, glassware—empty or filled with clear liquid	To emphasize surface irregularities	A
		To emphasize cracks, chips, and foreign particles	D or E

*A—Reflector spot lamps or well-shielded fluorescent lamps in a concentrating reflector.
B—Incandescent or mercury reflector without diffusing cover providing high luminous light.
C—A moderate-luminance source which includes most industrial-type fluorescent units having a variation in luminance of more than two to one.
D—A uniform-luminance-type unit such as an arrangement of lamps behind a diffusing panel.
E—A uniform-luminance-type unit superimposing stripes or lines.

time and energy and causes extra work for the eyes.

Color

In many industrial areas it is not necessary to distinguish colors with any significant degree of accuracy, and the appearance of the human complexion is less important than in many commercial areas. However, inspection and appraisal of certain materials, when done visually, require lighting that has been standardized in quantity and color quality. Where good color rendition is needed, incandescent, fluorescent, or color-improved mercury lamps are recommended. In other areas, when color is of little importance, clear mercury lamps can produce a very economical lighting system.

Where excellent color rendition is a special requirement, deluxe cool-white fluorescent lamps are usually the best single source.

For critical color matching, special equipment designed for the purpose is necessary.

Cost

In areas where the lighting is to be operated almost continuously, high-efficient sources, such as mercury or fluorescent lamps, are extremely desirable, since initial cost is of minor importance compared to operating cost. On the other hand, where the lamps are in operation for shorter periods of time, the initial cost is more significant, and incandescent lamps are recommended in spite of their lower efficiency. Power rate is another primary consideration in lighting economics. The higher the power rate, the higher equipment and lamp cost is justifiable, providing it results in a more efficient system and lower operating cost. A complete form for cost analysis is given in Unit 24.

LIGHTING INDUSTRIAL AREAS

Table 14-1 will act as a guide in selecting the proper light source for various industrial areas.

Industrial interiors generally fall into two classifications:

1. High-bay areas, where mounting height is 25 feet and higher.
2. Low-bay areas, where mounting height is 25 feet and lower.

High-bay lighting systems usually consist of mercury vapor or high-output (800 to 1500 mA) fluorescent fixtures, although high-pressure sodium lamps are finding their way into these areas. The lighting designer should keep in mind that economy factors are weighed against all other factors.

If mercury vapor lamps are the chief source for high-bay areas, they are usually used in combination with some incandescent lamps. The combination has definite advantages over a system using mercury lamps alone.

First, mercury lamps go out if there is even a momentary power interruption, and it may be 10 to 15 minutes before they will restart and come

PRINCIPLES OF ILLUMINATION

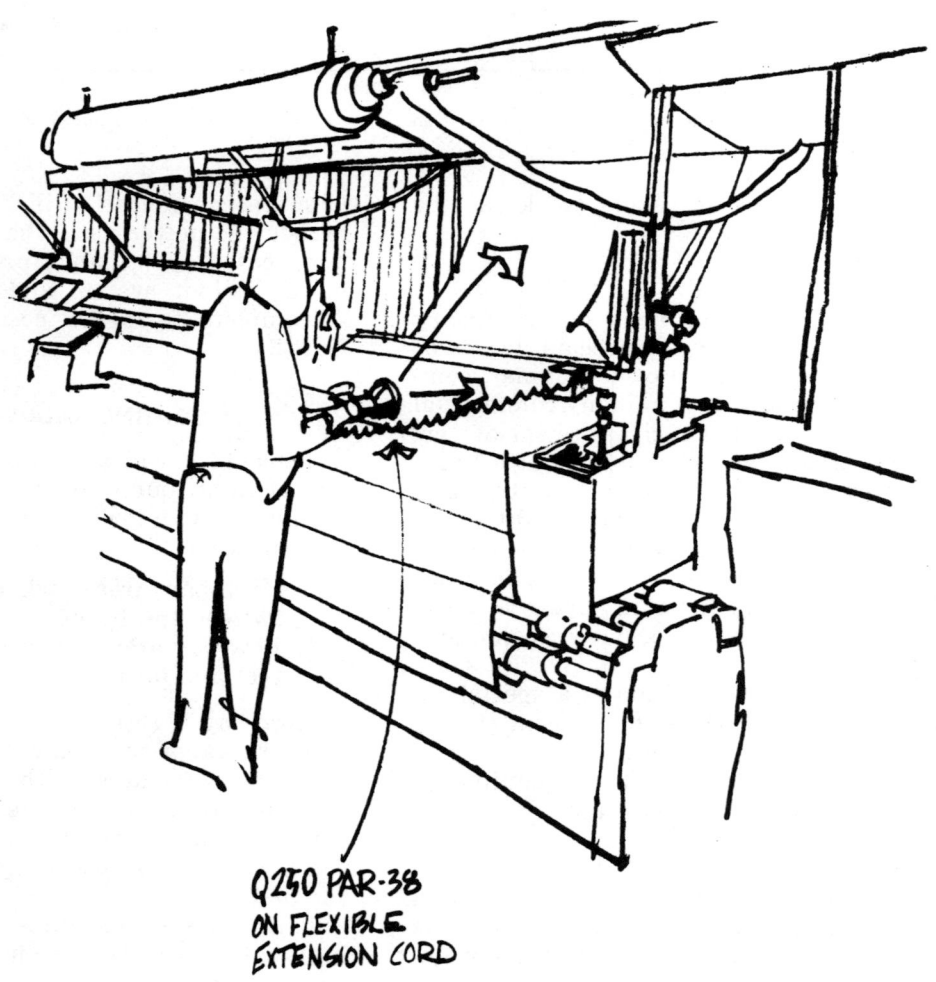

Fig. 14-1. The General Electric Q250 Par-38 quartzline lamps used in hand-held inspection lighting.

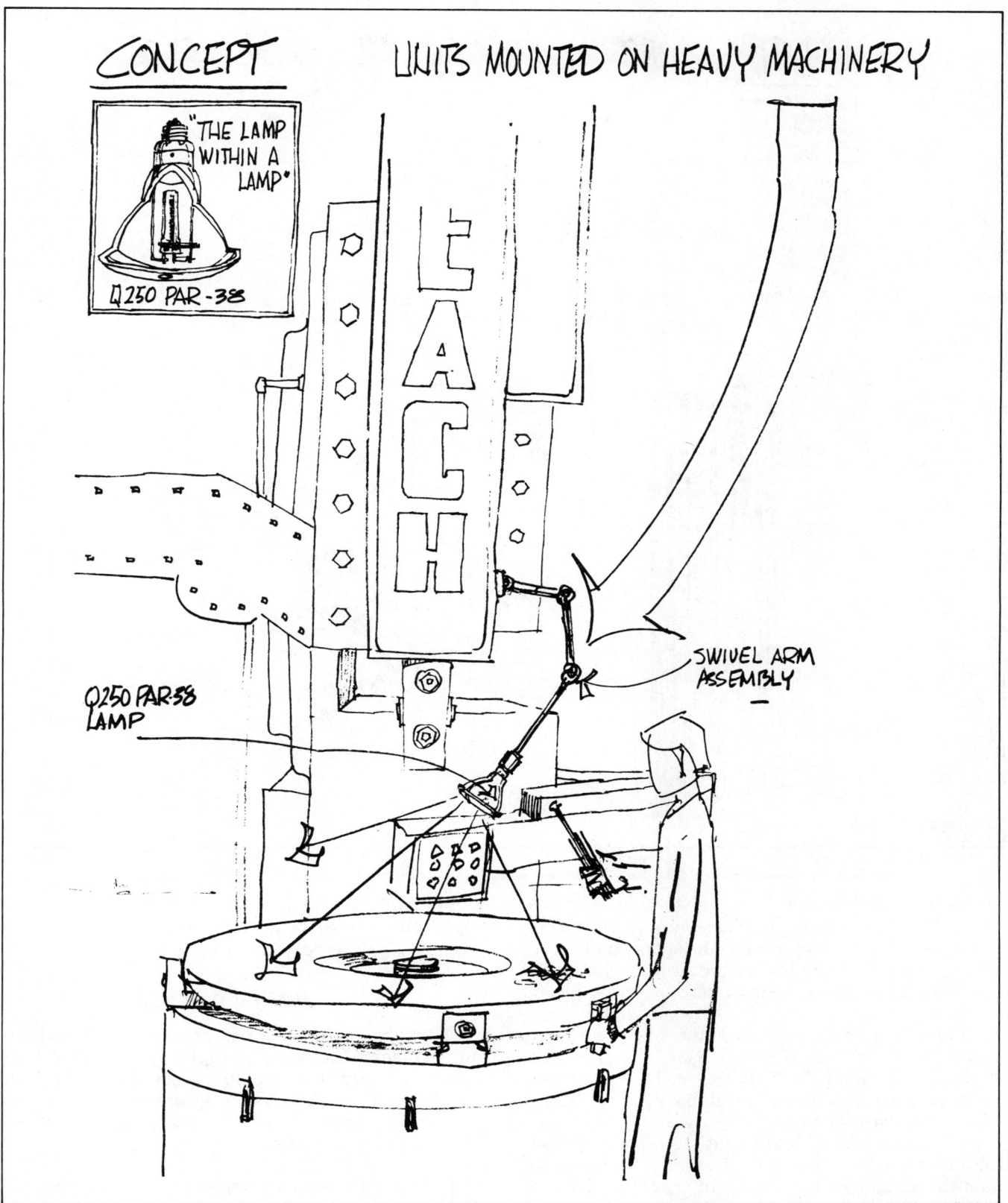

Fig. 14-2. Some General Electric Q250 Par-38 quartzline lamps mounted on heavy machinery.

PRINCIPLES OF ILLUMINATION

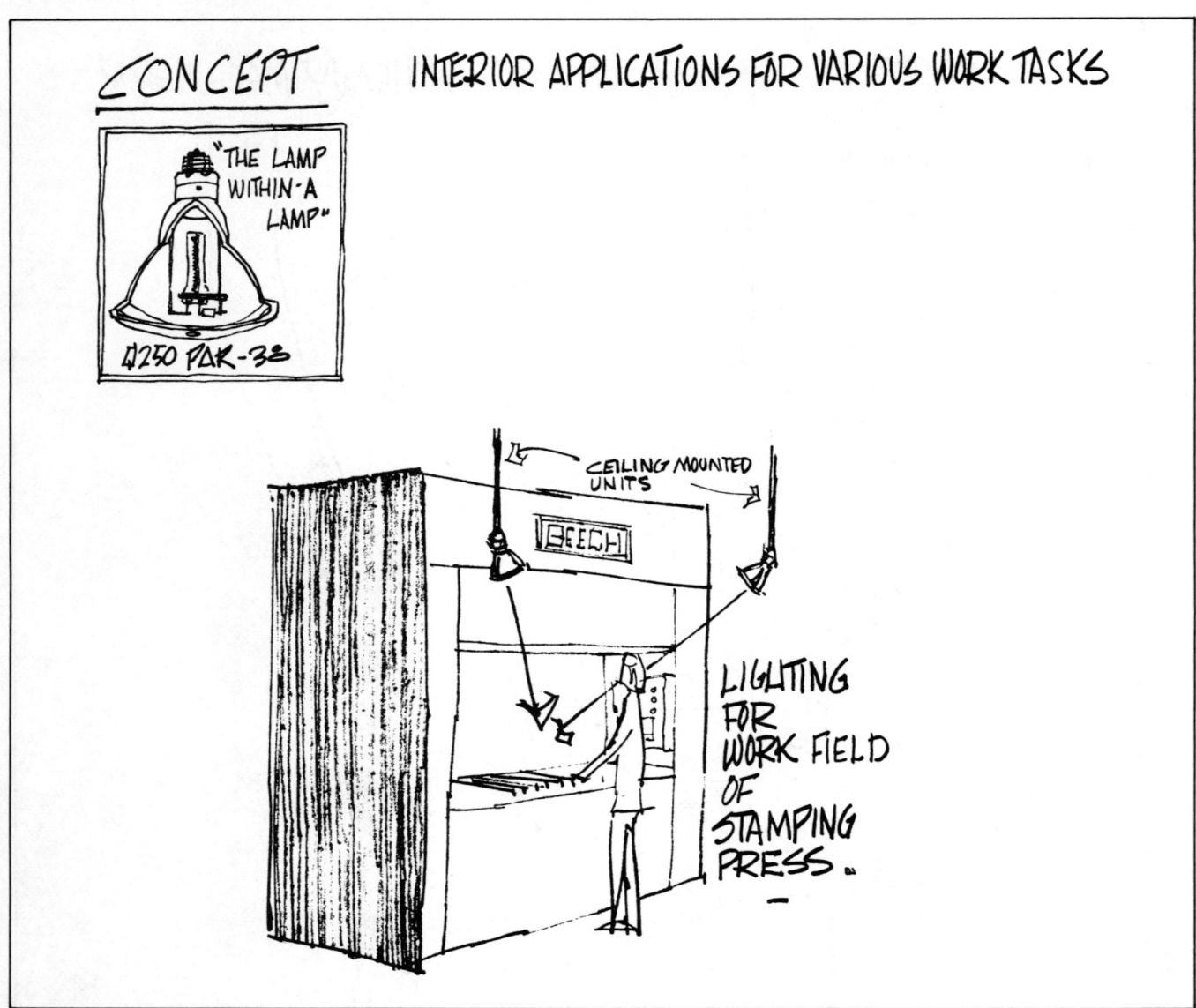

Fig. 14-3. The General Electric Q250 Par-38 quartzline lamps being used to light the work field of a stamping press.

up to full light output. This not only causes loss time for the workers, but such an interruption could be highly dangerous in some industrial areas. The light from the incandescent system will provide light should such a power interruption occur.

If the system is designed for approximately equal initial lumens for each type of lamp, the incandescent lamp will have to be of higher wattage than the mercury lamps. Quartz lamps would be a good choice.

In low-bay areas, fluorescent lighting is almost the universally accepted system. High-output fluorescent lamps that are eight feet long are the most economical and, therefore, are the ones normally used.

Industrial plants usually have office areas which should be treated in the same manner as the commercial lighting covered in Unit 11.

SUPPLEMENTARY LIGHTING

Supplementary lighting, as the name implies, supplements the general lighting by providing light for difficult seeing tasks or inspection processes. According to the specific situation, supplementary lighting may be installed for any of the following purposes:

1. For a directional controlled beam of light.
2. For additional properly shielded light with no special beam requirements.

INDUSTRIAL LIGHTING

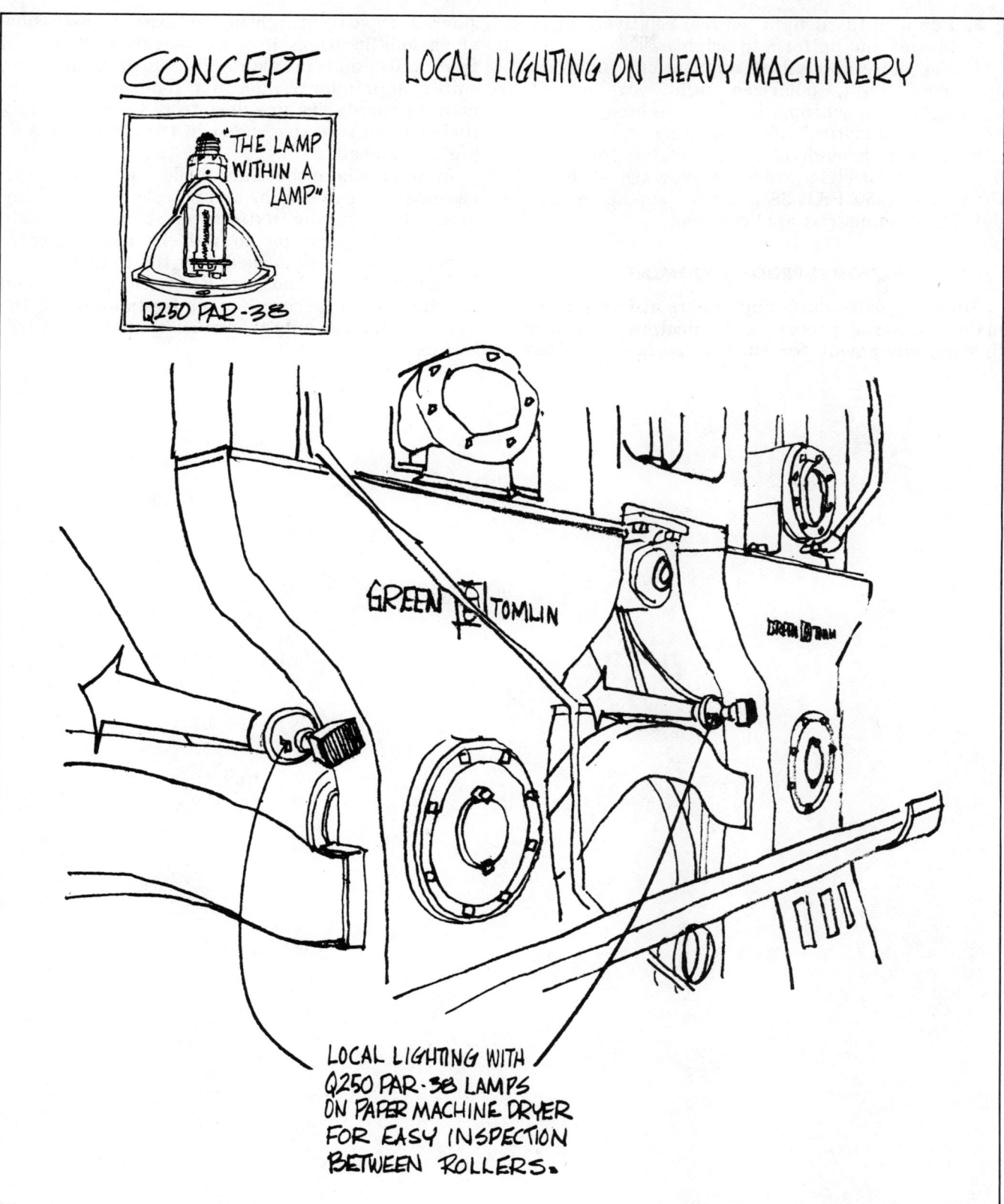

Fig. 14-4. Local lighting on heavy machinery with General Electric Q250 Par-38 quartzline lamps.

PRINCIPLES OF ILLUMINATION

3. For a diffused light source, relatively large in area and uniform in brightness.
4. For more specialized applications involving black light, polarized light, stroboscopic light, monochromatic light, lighted magnifier, and critical color matching.

Figs. 14-1 through 14-4 are sketches furnished by General Electric which show ways of using their new Q250 PAR-38 lamp for supplementary lighting in industrial applications.

EXPLOSIVE-PROOF EQUIPMENT

Since explosive dusts and vapors are present in many industrial processes, application of proper lighting equipment for such areas must be considered. Dust-tight lighting fixtures are necessary at various locations in grain elevators, flour and feed mills, and in plants in which cornstarch, sulfur, aluminum powder, and like substances are manufactured. The fine dust from these manufacturing processes, suspended in the atmosphere, is highly explosive.

In areas where there are explosive vapors, such as in paint spray booths and petroleum-processing areas, the lighting fixtures must be completely sealed with a gasket to prevent any explosive vapors from being sucked into the fixture.

Sections in the National Electrical Code should be studied for a complete understanding of the various requirements that must be met in hazardous locations.

SECTION V

EXTERIOR AND SPORTS LIGHTING

UNIT 15

General Floodlighting Design

The purpose of exterior floodlighting is to extend beyond sunset the usefulness and attractiveness of any given outdoor space or object. Whether the area is a sports field for amateur or professional engagements, a parking lot, shopping center, or back yard, properly applied lighting ensures safety, convenience, and a pleasing atmosphere.

This section has been prepared to cover basic design criteria for most outdoor lighting applications. Appendix A gives recommended levels of illumination in accordance with the latest Illuminating Engineering Society's recommended practice. Other tables have been prepared showing the types of equipment needed for the desired illumination and may be found in Units 15 through 19.

FLOODLIGHTING CALCULATIONS

Since floodlighting can encompass so many variations, and since the location of the floodlight relative to the object to be illuminated can be in any plane, any size, and any distance, standardization of design procedures is difficult. There are, however, certain fundamental laws of illumination which may be followed in designing floodlight installations.

The three most commonly used systems for floodlight calculations are the point-by-point method, the beam-lumen method, and the watts per square foot method. The point-by-point method, as covered in Unit 10, permits the determination of footcandles at any point and orientation on a surface. This method is valuable, since it permits a visualization of the degree of lighting uniformity realized for any given set of conditions. The beam-lumen method is very similar to the method covered in Unit 9 for interior lighting, except that it must take into consideration the fact that floodlights are not usually perpendicular to the seeing task, but instead are aimed at various angles to the surface.

BEAM-LUMEN METHOD

Beam lumens are defined as the amount of light that is contained within the beam spread of the floodlight. The lamp lumens, as found in lamp data tables, multiplied by the beam efficiency of the floodlight will give the beam lumens.

The coefficient of beam utilization (CBU), written as a decimal fraction, expresses the following ratio:

$$CBU = \frac{LA}{TBL}$$

where,
CBU is coefficient of beam utilization,
LA is lumens reaching area,
TBL is total beam lumens.

This is the percentage of the beam that falls on the area to be lighted. It can vary from 60 to 100% and can be accurately determined only through extensive calculations. Some guides can be established, however, which will allow a fairly accurate choice of beam utilization. In general, the larger the area to be illuminated, the higher the percentage of beam utilization.

The correct beam spread, too, is very important in obtaining the highest percentage of beam utilization. If the beam is wider than necessary, excessive light will be dispersed off the area and the beam utilization will be low. This is illustrated in Fig. 15-1.

Fig. 15-2 gives a comparison of approximate percentages of beam utilization factors and will serve for most floodlight applications.

PRINCIPLES OF ILLUMINATION

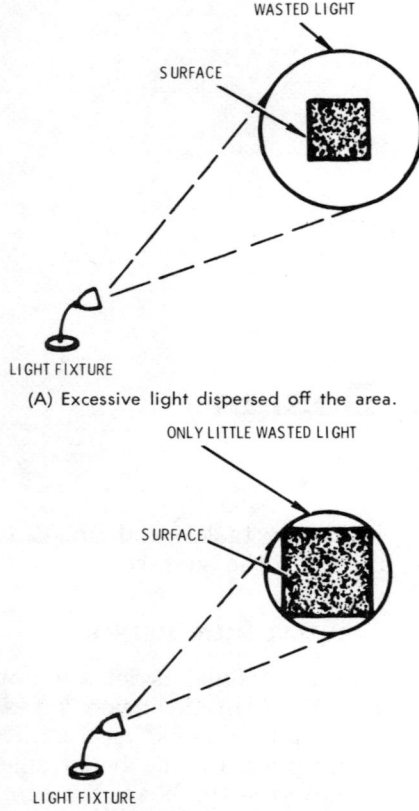

Fig. 15-1. Beam spread affects percentage of beam utilization.

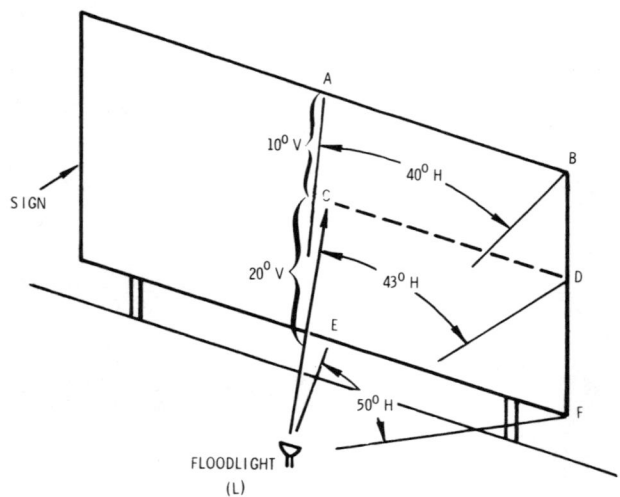

Fig. 15-3. A method of calculating beam utilization.

In Fig. 15-3 the floodlight is directed at point "C" on the sign and angle:

E(L)C = 20° Vertical
A(L)C = 10° Vertical
A(L)B = 40° Horizontal
C(L)D = 43° Horizontal
E(L)F = 50° Horizontal

Light distribution curves for most floodlights are available upon request. A typical curve is illustrated in Fig. 15-4.

All angles found by the floodlight in question to given points on the sign are then plotted on the grid of the isocandela curve. Because of the manner in which floodlights are photometered, all horizontal lines parallel to a line perpendicular to the beam axis appear as straight horizontal lines on the grid. All vertical lines through the beam axis appear slightly curved.

Therefore, all the lumens within the solid line ABEF fall on the sign. This totals 879 lumens and when doubled, to account for the other half of the beam, gives a total of 1758 lumens falling on the sign. Since the total beam lumens is 2748, and the total falling on the sign is 1758, we substitute these values in the formula:

$$\text{CBU} = \frac{1758}{2748}$$

$$= 0.64 \text{ or } 64\%$$

The total floodlights required for a desired level of illumination may be found by the formula:

$$\text{NF} = \frac{A \times DF}{BL \times CBU \times MF}$$

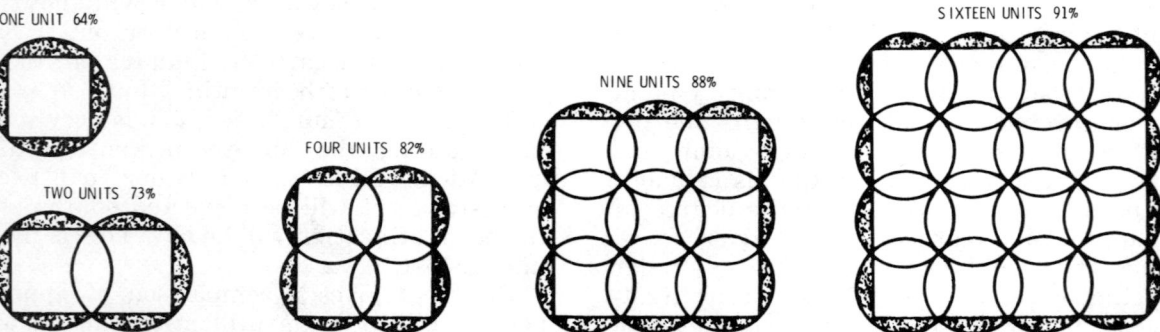

Fig. 15-2. Beam utilization comparison (percentages indicate beam utilization of floodlight).

GENERAL FLOODLIGHTING DESIGN

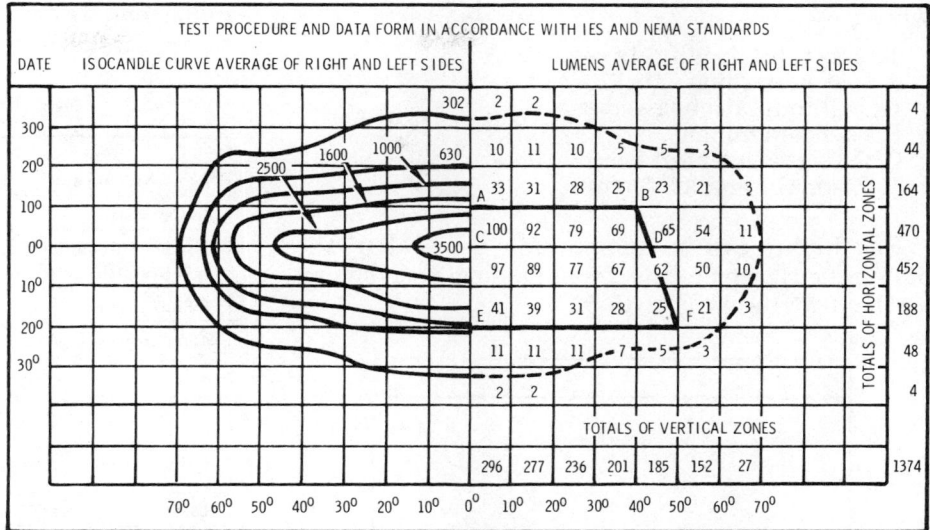

Fig. 15-4. Isocandela curve of 300 wide-beam reflector.

where,
NF is number of footcandles,
A is area,
DF is desired footcandles,
BL is beam lumens,
CBU is coefficient of beam utilization,
MF is maintenance factor.

The maintenance factor figure indicates that light output from floodlights drops off as they continue operating in service. One reason is because of dirt accumulation on the lens of the floodlight. This figure, of course, varies with the atmosphere in which the units are operating. Another reason is because of the drop-off in lumen output of the lamps as they operate throughout life. The light output from some lamps drops off just slightly, while others have a higher drop-off—see Section II and also lamp data catalogs. Table 15-1 lists the suggested total maintenance factors for various types of lamps.

Example Using Beam Lumen Method

An active storage area, 40 feet wide by 80 feet long, at an industrial plant is to be illuminated for

Table 15-1. Maintenance Factor for Various Lamps

Type of Lamp	Maintenance Factor
Incandescent	0.75
Quartz	0.85
Clear and Color-Improved Mercury	
175 to 700W	0.75
1000W	0.70
White Mercury	
175 to 700W	0.70
1000W	0.65
Metal Halide	0.65
Lucalox (sodium)	0.75

an average of 2 hours per night, 5 nights a week.

The system must be switched on and off several times during the night.

It is adjacent to a building 30 feet high.

Using the Beam Lumen Method, design the most practical lighting system.

Determine the Level of Illumination

The recommended illumination levels for many floodlighting applications are listed in Appendix A. This listing gives a recommended illumination level for an active storage yard at an industrial plant as 20 footcandles.

Determine the Type of Lamps to be Used

Due to the characteristics of the area, and especially due to the "switching" requirements, either heavy-duty incandescent or quartz lamps should be selected.

Determine the Type of Floodlight Fixture

A wide-beam floodlight which contains a 1500-watt quartz lamp is selected from the manufacturer's catalog. The floodlight beam lumens are also given in the catalog and are 23,338 for each fixture.

Determine the Coefficient of Beam Utilization and Maintenance Factor

By referring to Fig. 15-2, we can estimate the coefficient of beam utilization as 73% and a maintenance factor of 0.85 (from Table 15-1).

Therefore, the total number of floodlights may be found by substituting these values in the formula.

$$\text{TFL} = \frac{A \times IL}{BL \times CBU \times MF}$$

PRINCIPLES OF ILLUMINATION

where,

TFL is total floodlights required,
A is area in square feet,
IL is illumination level,
BL is beam lumens,
CBU is coefficient of beam utilization,
MF is maintenance factor.

Floodlights can be mounted on the roof of the building adjacent to the area and spaced as shown on the drawing in Fig. 15-5.

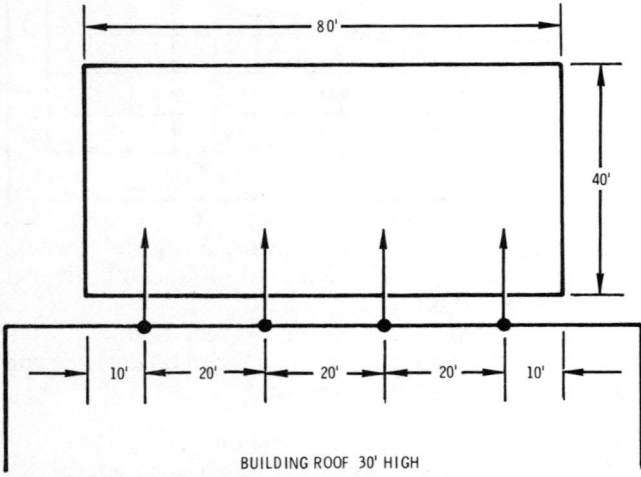

Fig. 15-5. Floodlights mounted on roof of building adjacent to the area to be illuminated.

WATTS PER SQUARE FOOT METHOD

For a quick and reasonably accurate way to determine the number of floodlights required for a lighting application under 10 footcandles, the "watts per square foot method" of calculation may be used. The formula is:

$$TFL = \frac{A \times IL \times WSF}{LWF}$$

where,

TFL is total floodlights required,
A is area in square feet,
IL is illumination level,
WSF is watts per square foot factor,
LWF is lamp watts per floodlight.

This formula eliminates the use of technical lighting terms in the calculations.

First, determine the proper illumination level from Appendix A.
Second, select the type of light.
Third, select type of floodlight by using manufacturers' catalogs to select the proper floodlight for the given application. Table 15-2 is listed in the Crouse-Hinds Outdoor Lighting and Selector guide and is typical of those found in manufacturers' catalogs and data.

Table 15-2. Light Source Selector

Characteristic	Incandescent	Quartz	Mercury	Metallic Additive	Fluorescent
Initial cost	Low	Low	Higher	Higher	Higher
Power consumption (for equal light)	Medium to High	Medium to High	Low	Low	Low
Fixture size	Medium	Small	Medium	Medium	Large
Long burning hours per year (over 1000)	Fair	Fair	Good	Good	Good
Short burning hours per year (under 1000)	Good	Good	Good	Good	Fair
Color definition	Good	Very Good	Fair	Good	Fair
Location considerations*	Fair	Fair	Good	Good	Fair
Beam control	Very Good	Good	Fair	Good	Poor
Cold weather operation	Very Good	Very Good	Good	Good	Fair
Long-range projection (narrow beam)	Best	Fair	Fair	Fair	Poor
Medium-range projection	Good	Good	Good	Good	Fair
Annual operating cost	Medium	Medium	Low	Low	Low

*Fixtures difficult or expensive to relamp and service.

Fourth, determine the number and the placement of floodlights and poles.

In the watts per square foot formula, lamp lumens, floodlight beam efficiency, and lumen maintenance factors are all combined in an estimated utilization factor which will be known as the WSF factor. Four such factors are required in order to provide reasonable accuracy in different size areas. These are shown in Table 15-3.

Table 15-3. WSF Factor for Different Areas

Area	WSF
Small area (1000 to 3000 sq ft)	0.16
Medium area (3000 to 20,000 sq ft)	0.11
Large area (20,000 to 80,000 sq ft)	0.08
Extra large area (over 80,000 sq ft)	0.06

For example, using the watts per square foot formula, determine the quantity, wattage, and positioning of floodlights required to light a shopping center parking area which is 240 feet wide and 420 feet long. The floodlights will operate approximately 2000 hours per year.

Appendix A recommends an illumination level of two footcandles. We will use this figure for our design.

From the data in Table 15-2 we concluded that either mercury or metallic additive lamps would

GENERAL FLOODLIGHTING DESIGN

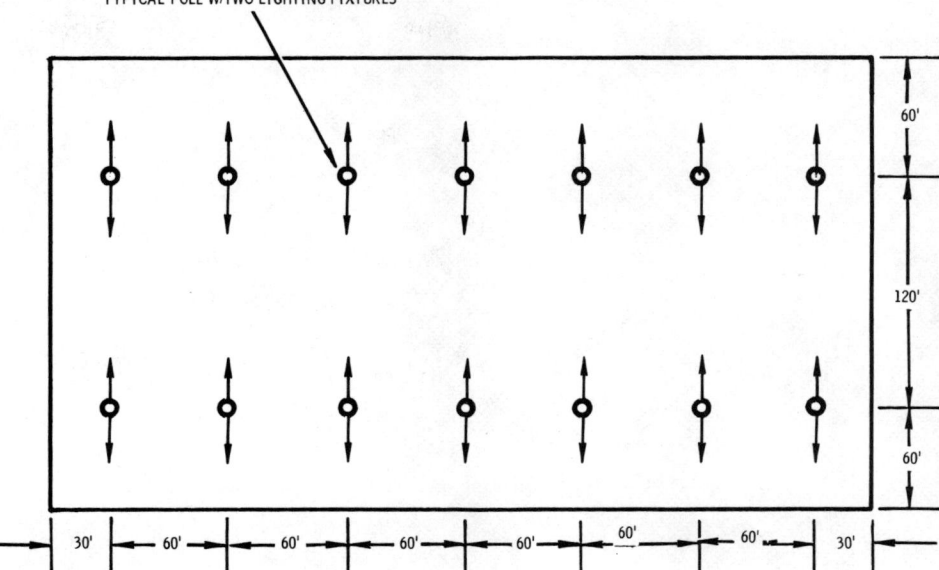

Fig. 15-6. Pole and fixture layout for parking lot.

be the best choice. For this application, use the metallic additive type.

We will choose a suitable lighting unit which will handle a 400-watt metallic additive type light.

By substituting our known value in the watts per square foot formula we have:

$$\text{TFL} = \frac{90,800 \times 2 \times 0.06}{400}$$
$$= 27.2 \text{ floodlights}$$

Using 28 floodlights, they may be laid out as shown in Fig. 15-6.

UNIT 16

Sports Lighting

The interest in both outdoor and indoor sports is international in scope, and the lighting of recreational areas for night use permits thousands of people, who are occupied during the day, to view and to take part in sports.

In designing lighting for sports, careful consideration should be given to the requirements and comfort of each of three groups: players, officials, and spectators. Providing adequate illumination for one group should not introduce objectionable glare into the field of view of the other two groups. Lighting systems for areas where sports will be televised usually requires a higher level of illumination and also a higher quality of light. This not only ensures best color broadcast rendition, but vastly improves visibility for both spectators and participants.

LIGHTING FOR INDOOR SPORTS

The design and calculation procedures for interior lighting are outlined in Units 8 and 9. Recommended illumination levels for various indoor sports my be found in Appendix A. However, in addition to these procedures, it is necessary in designing sports lighting to consider the following factors:

1. Spectators and participants have no fixed field of view. During the course of the game (basketball is a good example), the ceiling and luminaires may frequently be included in the visual field.
2. The object of regard, such as a basketball, will have no fixed location and may be viewed at floor level, near the ceiling, or at any level in between.
3. It is particularly important for participants to be able to accurately estimate the speed and line of flight of the object.
4. If the area is to be used for just one sport, then the lighting design problems are easier. However, many areas, such as a field house, may be used for more than one sport, and the lighting system must meet the varied or particular requirements for each sport.

Table 16-1. Recommended Floodlight Arrangement for Baseball Fields

Class	Recommended Footcandles		Floodlights		Min. Mtg. Ht.[1] (ft.)
	Infield	Outfield	Type	Total No.[2]	
Major League	150	100	3, 4, or 5	1000	120
AAA and AA	70	50	3, 4, or 5	500	110
A and B	50	30	3, 4, or 5	320	90
C and D	30	20	3, 4, or 5 or 4, 5, or 6	240 320	70
Semipro and Municipal	20	15	3, 4, or 5 or 4, 5, or 6	160 220	70
Recreational	15	10	3, 4, or 5 or 4, 5, or 6	120 160	70

[1] To bottom floodlight crossarm.
[2] Based on 1500-watt PS-52 general service filament lamps operated at 10% overvoltage 45,000 lumens.

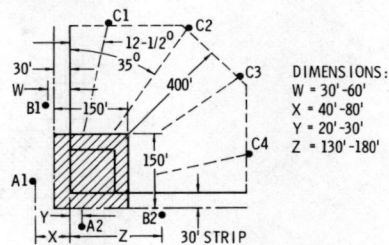

PRINCIPLES OF ILLUMINATION

Table 16-2. Recommended Arrangement for Junior-League, Class I Baseball Fields

League	Recommended Illumination Levels		Pole Data		VRC-18 (Incandescent)			MRF-400			Quartz			MFB-1000		
	Infield	Outfield	Mounting Height (ft)	Location	Total Per Pole	NB	WB	Total Per Pole	NB	WB	Total Per Pole	NB	WB	Total Per Pole	NB	WB
Junior League Class I	30	20	40	A_1, A_2	6	..	12	12	..	24	8	..	16	6	..	12
			40	B_1, B_2	12	..	24	24	..	48	10	..	20	12	..	24
			50	C_1, C_2, C_3, C_4	6	..	24	12	..	48	4	..	16	6	..	12
					..	..	60	..	..	120	..	..	52	..	..	48

Note: All lighting fixture numbers (VRC-8, MRF-400, etc.) are as designated by Westinghouse Electric.

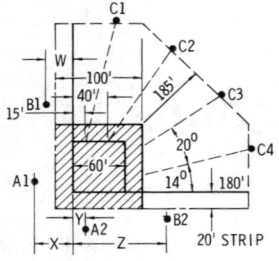

THESE RECOMMENDATIONS ARE BASED UPON THE FOLLOWING:

Total playing area, including a strip 20' outside of each foul-line . . . . . 34,700 sq ft
Infield area (shaded) . . . . . . 10,000 sq ft
Outfield area . . . . 24,700 sq ft

DIMENSIONS
W = 20'-30'
X = 30'-50'
Y = 5'-15'
Z = 90'-110'

LIGHTING FOR OUTDOOR SPORTS

Baseball Fields

The basic requirements for a good installation of night baseball lighting are:

1. Adequate intensity to meet the demands of the particular class of play and the maximum spectator viewing distance.
2. Correct distribution and aiming of the floodlights to ensure best utilization of the light with maximum ease of sight for players, spectators, and officials.
3. Uniform distribution of the light on the ground and in the air.

Due to the comparatively long distance that the light must be projected in baseball lighting, some high-candlepower floodlights are required, regardless of the intensity desired.

Many fields may be successful lighted by a strict adherence to the layout in Table 16-1. Others require modification to meet local conditions, such as odd field size or shape, limitations on pole locations, unusual stand construction or size, and unusual distance between base lines and stands. Whatever modification of the basic layouts or distribution is required should be determined by a careful study of the problem. It is recommended that planning aids be obtained from experienced outdoor lighting manufacturers. Standard floodlighting arrangement for Junior League Class I

Table 16-3. Recommended Arrangement for Junior-League, Class II Baseball Fields

League	Recommended Illumination Level		Pole Data		VRC-18 (Incandescent)			MFB-1000			Quartz			MLS-20 (1000 W. Met. Hal.)	
	Infield	Outfield	Mounting Height (ft)	Location	Total Per Pole	NB	WB	Total Per Pole	NB	WB	Total Per Pole	NB	WB	Total Per Pole	WB
Junior League Class II	40	30	50	A_1, A_2	14	12	16	8	..	16	16	14	18	8	16
			50	B_1, B_2	24	24	24	14	..	28	18	14	22	14	28
			50	C_1, C_2, C_3, C_4	14	36	20	9	..	36	9	12	24	8	32
					132	72	60	80	..	80	104	40	64	76	76

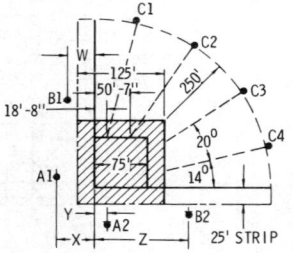

THESE RECOMMENDATIONS ARE BASED UPON THE FOLLOWING:

Total playing area, including a strip 25' outside of each foul-line . . . . . 62,225 sq ft
Infield area (shaded) . . . . . . 15,625 sq ft
Outfield area . . . . 46,600 sq ft

DIMENSIONS:
W = 25'-45'
X = 35'-65'
Y = 10'-25'
Z = 110'-145'

Table 16-4. Recommended Arrangements for Illuminating Softball Fields

Class of Play	Maintained Illumination		Pole Data		VRC-18—1500 Watt			Quartz—1500 Watt			1500 W. Mercury MFB-1000 Total Per Pole	1000 W. MH Lamp MLS Floods		
	Infield	Outfield	Mounting Height (ft)	Location	Total Per Pole	NB	WB	Total Per Pole	NB	WB		Total Per Pole	NB	WB
Professional & Championship 280' Outfield	50	30	50	A_1, A_2	14	4	10	12	4	8	9	8	2	6
			50	B_1, B_2	30	16	14	23	10	13	18	18	9	9
			60	C_1, C_2, C_3, C_4	18	6	12	14	7	7	11	11	5	6
				Total	160	64	96	126	56	70	98	96	42	54
Semipro 280' Outfield	30	20	40	A_1, A_2	8	..	8	6	..	6	5	5	..	5
			40	B_1, B_2	18	9	9	15	7	8	10	11	5	6
			55	C_1, C_2, C_3, C_4	14	4	10	12	5	7	9	8	3	5
				Total	108	34	74	90	34	56	68	64	22	42
Semipro 240' Outfield	30	20	40	A_1, A_2	8	..	8	6	..	6	5	5	..	5
			40	B_1, B_2	14	6	8	13	5	8	9	8	3	5
			50	C_1, C_2, C_3, C_4	10	3	7	9	4	5	7	6	2	4
				Total	84	24	60	74	26	48	56	50	14	36
Industrial 280' Outfield	20	15	35	A_1, A_2	6	..	6	5	..	5	4	4	..	4
			35	B_1, B_2	14	6	8	11	3	8	8	8	3	5
			50	C_1, C_2	20	6	14	17	5	12	12	12	4	8
				Total	80	24	56	66	16	50	48	48	14	34
Industrial 240' Outfield	20	15	35	A_1, A_2	6	..	6	5	..	5	4	4	..	4
			35	B_1, B_2	10	3	7	8	..	8	7	6	2	4
			45	C_1, C_2	14	4	10	11	..	8	8	8	3	5
				Totals	60	14	46	48	..	48	38	36	10	26
Industrial 200' Outfield	20	15	35	A_1, A_2	5	..	5	4	..	4	4	3	..	3
			35	B_1, B_2	7	..	7	5	..	5	5	5	..	5
			40	C_1, C_2	10	2	8	8	..	8	5	6	2	4
				Total	44	4	40	34	..	34	28	28	4	24
Recreational 200' Outfield	10	7.5	35	A_1, A_2	3	..	3	3	..	3	2	2	..	2
			35	B_1, B_2	4	..	4	3	..	3	3	3	..	3
			40	C_1, C_2	5	..	5	4	..	4	3	3	..	3
				Total	24	..	24	20	..	20	16	16	..	16

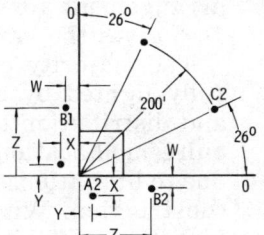

DIMENSIONS:
W = 20'-30'   Y = 5'-15'
X = 25'-50'   Z = 90'-110'

baseball fields is shown in Table 16-2. See Table 16-3 for the floodlighting arrangement of Junior League Class II baseball fields.

Softball Fields

The basic requirements for a good installation of night softball lighting are the same as for night baseball. Table 16-4 gives the layout for various classes of play. Many fields may be successfully lighted by a strict adherence to this layout and table.

Football Fields

It is generally conceded that distance between the spectators and the play is the first consideration in determining the class and lighting requirements. However, the potential seating capacity of the stands should also be considered.

The pole plans, as indicated in Table 16-5, are considered good practice, with local field conditions dictating the exact pole locations.

The basic requirements for a good installation of night football lighting are the same as for night baseball. The standards recommended by Illuminating Engineering Society, if adhered to, ensure the correct number of floodlights and correct pole locations. The most important variable in football floodlighting is the distance from the poles to the sidelines. As this distance increases, it is necessary to increase the mounting height and to use flood-

PRINCIPLES OF ILLUMINATION

Table 16-5. Football Field Lighting Arrangement

IES Class of Play*	Maintained Illumination (Footcandles)	Pole Data			VRC-18 15,000 Watt			Quartz, 1500 Watt			1000W MH Lamp MLS-Floods		
		Setback (ft)	Mounting Height (ft)	Pole Quantity	Total Per Pole	NB Per Pole	WB Per Pole	Total Per Pole	NB Per Pole	WB Per Pole	Total Per Pole	NB Per Pole	WB Per Pole
II	50	75-100	75-100	6	36 .. Total 216	36 .. 216	.. .. ..	28 .. 168	21 .. 126	7 .. 42	22 .. 132	22 .. 132	.. .. ..
		50-75	50-75	8 6 Total	24 .. 192	24 .. 192	.. .. ..	.. 26 156	.. 14 84	.. 12 72	15 .. 120	15 .. 120	.. .. ..
III	30	30-50	30-50	8 Total	16 128	.. ..	16 128	12 96	.. ..	12 96	10 80	.. ..	10 80
IV	20	15-30	25-30	8 10	.. 8 80	.. .. ..	.. 8 80	8 .. 64	.. .. ..	8 .. 64	5 .. 40	.. .. ..	5 .. 40

*See Table 16-6.

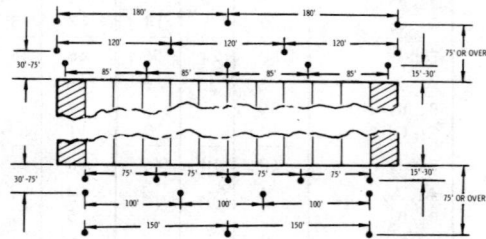

lights with higher peak candlepower and a narrower beam spread. The use of floodlights with too narrow a beam spread will result in spotty illumination. The use of floodlights with too wide a beam spread will result in poor utilization and a low level of illumination, particularly in the center of the field, and may produce glare in opposite stands close to the sidelines. Table 16-5 indicates a good arrangement for normal football lighting. Table 16-6 gives the various classifications for fields.

The majority of football fields may be successfully lighted by a strict adherence to the layout and distribution table. Some installations may require modification of the standard lighting plans due to limitations on pole location or unusual stand construction. Whatever modification of the basic layouts or distribution is required should be determined by a careful study of the problem.

Table 16-6. Football Field Classifications

Class	Distance (ft)	Seating Capacity	Footcandles Maintained
I	100-140	Over 30,000	100
II	50-100	10,000-30,000	50
III	30-50	5,000-10,000	30
IV	15-30	5000	20
V	†	†	10

*Distance from nearest sideline to farthest row of spectators.
†No fixed seating facilities.

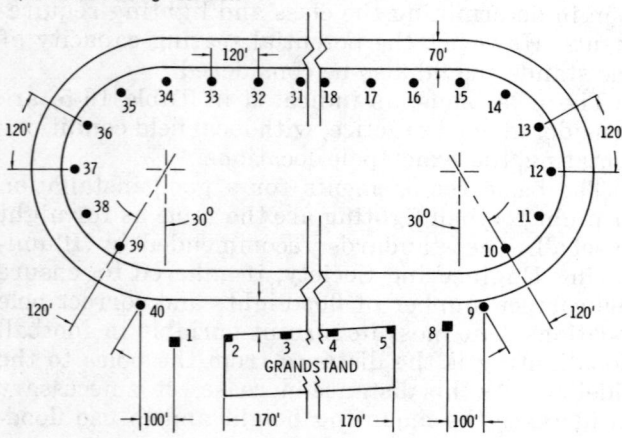

One-Mile Horse Race Track

In planning a lighting installation for a race track, there are two major considerations. First, the provision of light of adequate quality and quantity to permit vision by spectators, jockeys, horses, and officials.

Second, the appearance of the track must be good both at night and in the day. The latter factor is particularly important if, to secure maximum utilization of the plant, racing dates are scheduled both for night and day events.

The use of relatively few steel poles and underground wiring provides for the most attractive

SPORTS LIGHTING

Table 16-7. Recommended Arrangement of Lighting for Horse Race Tracks

Pole Data			Illumination Level Maintained	VRC-18 NP Inc. 1500 Watt			1000W MH Lamp MLS-20 Flood			Quartz 1500 W Lamp			MFB-1000 1500 W. Color Corrected	
Locations	Mounting Height (ft)	Quantity of Poles		Total Per Pole	NB Per Pole	WB Per Pole	Total Per Pole	NB Per Pole	WB Per Pole	Total Per Pole	NB Per Pole	WB Per Pole	Total Per Pole	WB Per Pole
1-6; 11-12	80[1]	8		30	24	6	19	15	4	24	19	5	20	20
7-10	100[2]	4		30	30	..	18	18	..	23	23	..	20	20
13 & 32	60[3]	2	20 Ftc.	28	22	6	18	14	4	22	17	5	19	19
14 & 31	60[3]	2		26	20	6	16	12	4	21	16	5	18	18
15-30	60[3]	16		24	18	6	15	11	4	19	14	5	16	16
			Totals	852	684	168	532	420	112	674	534	140	570	570
1-8	40[4]	8	20 Ftc.	23	15	8	14	9	5	18	12	6		
9-40	[5]	32		16	11	5	10	7	3	12	9	3		
			Totals	696	472	224	432	296	136	528	384	144		

[1] Poles set back 50 ft. from track.
[2] Poles set back 75 ft to 100 ft from track.
[3] Poles set back 40 ft from track.
[4] Poles set back 40 ft from track. Mounting height to be not less than 40 ft.
[5] Poles set back dependent upon location of grand stand. Projection distance "C" may dictate use of different beam spreads.

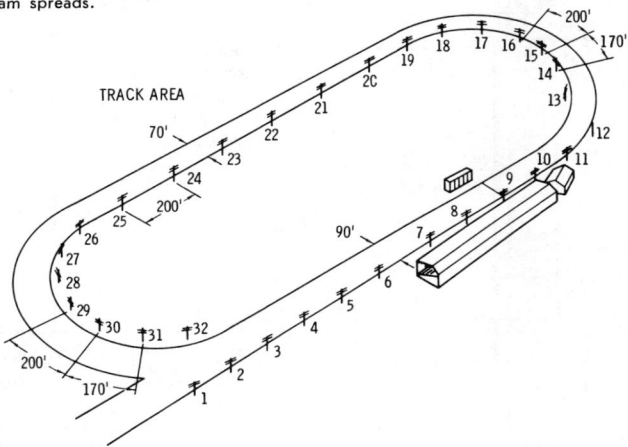

Table 16-8. Lighting Arrangement for One-Fourth Mile Stock Car Race Track

Pole Data			Illumination Level Maintained	VRC-18 Inc. 1500 Watt			1000 Watt MH MLS-20 Floods			Quartz 1500 W.	
Locations	Mounting Height (ft)	Quantity		Total Per Pole	NB Per Pole	WB Per Pole	Total Per Pole	NB Per Pole	WB Per Pole	Total Per Pole	MB Per Pole
1-3[1]	60	3	20	16	11	5	10	7	3	13	13
4-9[2]	60	6	20	16	16	..	10	10	..	13	13
			Totals	144	129	15	90	81	9	117	117

[1] Poles 1-3 are set back 25 ft from track.
[2] Poles 4-9 are set back 60 ft from track.

PRINCIPLES OF ILLUMINATION

Table 16-9. Lighting Arrangement for a Drag Strip

Area	Length (ft)	Illumination Values Footcandles	
		Initial	Maint
Staging	500	24	20
Acceleration	1320	24	20
Deceleration	1320	24	20
Shutdown	860	12	12
Return	...	Spill	Spill
Total	4000		

Pole	No. Poles	Quartz-Floods Narrow Beam	
		Each Pole	Total
A	15	14	210
B	4	7	28
Total	19	..	238
Total Connected Load ....414 kW[1]			

[1] 1500-watt quartz lamp 1500/CL operated at 10% over voltage.

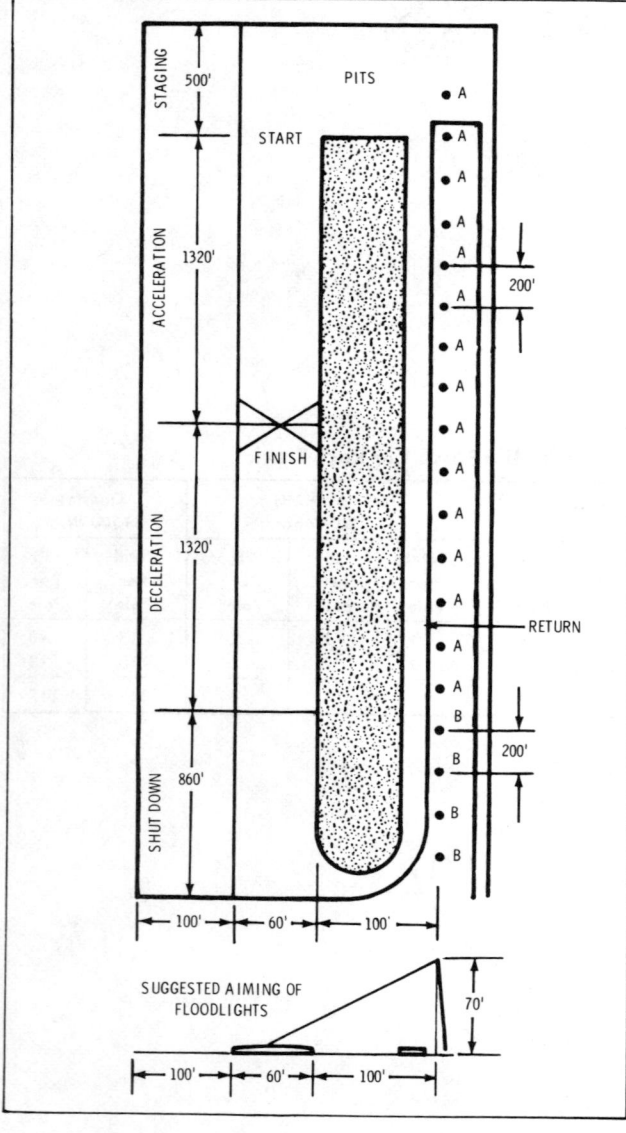

possible appearance (Table 16-7). The correct location of these poles, the correct mounting height, and the proper beam spread floodlights will help to achieve a lighting installation of high uniform intensity and complete freedom from glare. This will permit the use of even poor quality binoculars for following the race. It should be noted that, for those poles located on the turns and backstretch, the crossarms are oriented to direct the majority of the light in the direction of the race, thus ensuring a minimum of glare for jockeys and horses.

Table 16-10. Floodlighting for Single Tennis Courts

Class of Play	Illumination Footcandles		Floodlights Per Pole			Total Load (kW)
	Initial	Maintained	A	B	Total	
Using 1500-Watt Quartz Floodlights						
Recreation[1] (4 Poles)	12	10	1	..	4	6
Club (8 Poles)	24	20	1	1	8	12
Tournament (8 Poles)	36	30	2	1	12	18
Using 1000-Watt Mercury Floodlights						
Recreation (4 Poles)	15	10	2	..	8	8.8[2]
Club (8 Poles)	30	20	2	1	12	13.2[2]
Tournament (8 Poles)	45	30	3	2	20	22[2]

[1] For recreation play only, 4 poles may be used at A location except spaced 60 ft apart.
[2] Includes ballast losses. Minimum mounting height 30 ft.

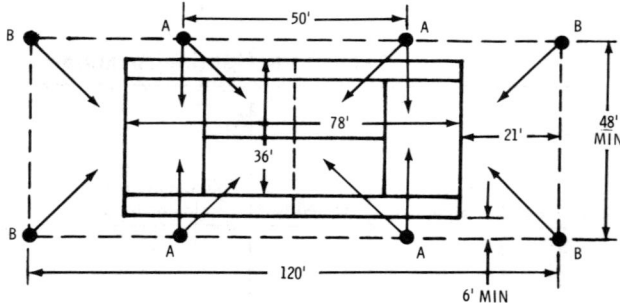

One-Fourth Mile Stock Car Track

For automotive racing, the visual task for spectator and driver is quite different, with the most difficult visual task being that of the driver. Adequate intensities of the order of 20 footcandles or more are thus imperative.

With spectators located on one side of the track, as shown in Table 16-8 it is essential that poles be placed in the track infield to ensure adequate light on the vertical surface seen by the spectators. In this layout, the poles have been located back far enough so that the driver can pull into the infield, in case of trouble, without danger of hitting one of the poles.

Table 16-11. Floodlighting for Two Tennis Courts

Class of Play	Illumination Footcandles		Number of Floodlights Per Pole			Total Load (kW)
	Initial	Maintained	A	B	Total	
Using 1500-Watt Quartz Floodlights						
Recreation (4 Poles)	12	10	2	..	8	12
Club (8 Poles)	24	20	3	1	16	24
Tournament (8 Poles)	36	30	4	2	24	36
Using 1000-Watt Mercury Floodlights						
Recreation (4 Poles)	15	10	2	..	8	8.8[2]
Club (8 Poles)	30	20	3	1	16	17.6[2]
Tournament (8 Poles)	45	30	4	2	24	26.4[2]

[2] Includes ballast losses. Minimum mounting height 30 ft.

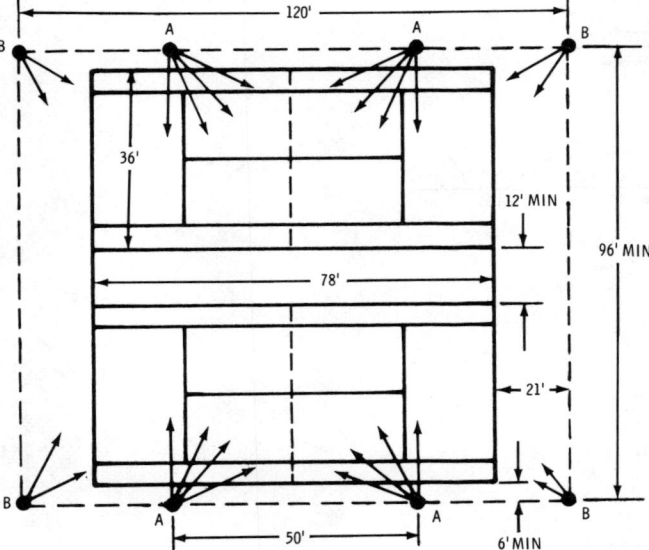

Drag Strip Lighting

Drag strip application requires adequate lighting for both the participants and the spectators. The lighting must also be uniform and glare-free. The lighting arrangement of a drag strip is shown in Table 16-9.

Tennis Court Lighting

In designing lighting for tennis courts, careful consideration should be given to the requirements and comfort of the players, officials, and spectators. The basic requirements for a good installation of night tennis court lighting are:

1. Adequate intensity to meet the demands of the players and the maximum spectator viewing distance.
2. Correct distribution and aiming of the floodlights to ensure best utilization of the light with maximum ease of sight for both players and spectators.
3. Uniform distribution of the light on the ground and in the air.

The recommended arrangement of floodlights for various classes of play on one tennis court is illustrated in Table 16-10. See Table 16-11 for the recommended arrangement of floodlights for various classes of play on two tennis courts.

Driving-Range and Miniature Golf

For a golf driving-range, the same lighting layout may be used whether the tees are in a straight line or in an arc. See Table 16-12 for the recommended arrangement of floodlights for golf driving-ranges.

Miniature Golf Area Lighting

Lighting should be designed to provide an average of 10 footcandles of horizontal illumination maintained in service.

Two basic methods may be used and are described as follows:

1. Floodlighting poles should be located around the perimeter of the area with a recommended mounting height of 20 to 25 feet. Spacing between the poles should not exceed two times the mounting height. One 1000-watt mercury lamp floodlight per pole should be used. Two 1500-watt clear incandescent lamps per pole or 2 quartz wide-beam floodlights with 1500-watt lamps could also be used.
2. The second method uses large area types of units with symmetric distribution on poles in the playing area. The recommended mounting height is 25 to 30 feet. Spacing between the poles should not exceed two times the mounting height.

Table 16-12. Arrangement of Floodlights for a Golf Driving-Range

Recommended Footcandles Maintained	Floodlights		Floodlight Recommended Per Pole Location			
	Aiming Point	IES Type	1500W VRC-18	1500 W Quartz	1000 W Mercury MFB-10	1000 W Mercury AEC-20
10 FC on tee	X	5 or 6	1-WB	1-WB	1-WB[1]	....
5 FC on vertical surface at 200 yards (approximately 180 meters)	Y	3	2-NB	....	....	2-NB[2]
	Z	2 or 3	3-NB	....	....	3-NB[2]

[1] Mercury on the tee to use high output white or metal halide lamps.
[2] Use clear mercury or metal halide lamps.

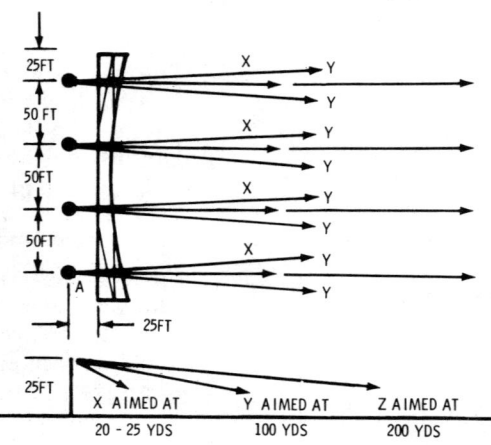

UNIT 17

Underwater Lighting

This unit will discuss the fundamentals of designing underwater lighting for fountains and swimming pools. Such lighting is used mainly for decorative purposes and can be compared to a painting or any other work of art. While the designer's artistic ability is very important in lighting designs of this type, there are certain basic rules which may be followed in order to produce good lighting layouts.

FOUNTAIN LIGHTING DESIGN

A fountain is utilized for one or more of the following reasons:

1. Sheer fascination of visual and sound effects.
2. To create product or trademark identification.
3. A decorative feature piece.
4. For animation.
5. Air conditioning reject heat load.
6. Enhance the surroundings of outstanding architectural structures.

Considerations for fountain lighting should include:

1. The type of water effect to be lighted.
2. Color selection.
3. Maximum height to be illuminated.
4. Selecting the type and number of fixtures and lamps.

TYPES OF WATER EFFECT TO BE LIGHTED

Single-nozzle fountains are normally lighted with two fixtures with spot lamps, such as illustrated in Fig. 17-1.

The smaller spray ring fountains may be lighted with one fixture, such as illustrated in

UNDERWATER LIGHT FIXTURE

Fig. 17-1. A single nozzle fountain.

Fig. 17-2. The larger spray ring fountains are normally lighted with one fixture every 4 to 5 feet for illuminating heights up to 15 feet. Wide-angle flood lamps are used. For heights over 15 feet, use one fixture every 2 to 3 feet with medium-spread flood lamps. See example in Fig. 17-3. This procedure also applies to waterfalls and weirs.

In all of the cases above, the lens of the fixture should be submerged two inches below the water level for proper operation.

PRINCIPLES OF ILLUMINATION

Fig. 17-2. A spray ring fountain.

Color Selection

The selection of colors is a subjective matter and can vary as much as opinions of designers or owners of buildings. However, the following may be used as a guide in selecting colors of lamps.

Colors directly affect the selection of fixtures and lamps, inasmuch as various colors require differing candlepower to achieve decorative effects of light.

Amber and Turquoise lamps require 50% more candlepower than a clear lens for the same level of illumination.

Red lamps require 100% more candlepower than a clear lens for the same level of illumination.

Blue and Green lamps require 250% more candlepower than a clear lens for the same level of illumination.

Where high levels of illumination surround the fountain, use caution when selecting colors; the surrounding light will tend to wash out the colored

Table 17-1. Minimum Desirable Beam Candlepower

Height of Water Effect (ft)	Clear	Amber & Turquoise	Red	Blue & Green
5	4,000	6,000	8,000	14,000
10	11,000	16,000	22,000	38,000
15	21,000	31,000	42,000	73,000
20	34,000	51,000	68,000	119,000
25	50,000	75,000	100,000	175,000
30	69,000	103,000	138,000	241,000
35	91,000	136,000	182,000	378,000
40	115,000	174,000	230,000	406,000
45	144,000	216,000	288,000	504,000
50	170,000	256,000	340,000	595,000

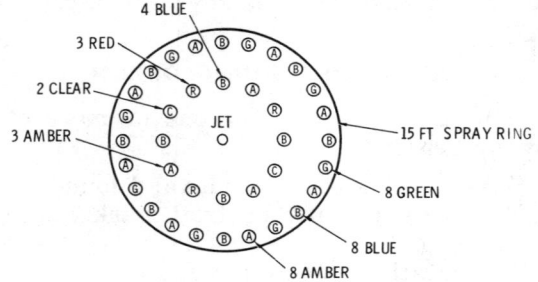

Fig. 17-3. A large spray ring fountain.

Table 17-2. Standard Lamps and Their Rated Candlepower

Watts	Bulb	Ordering Code	Beam Type	Beam Spread (degrees)	Initial Average Maximum Beam Candlepower	Approx Life (hours)
150	PAR-38	150PAR/SP	Spot	30X30	10,500	2000
150	PAR-38	150PAR/FL	Flood	60X60	3500	2000
150	R-40	150R/SP	Spot	40	6300	2000
150	R-40	150R/FL	Flood	110	1300	2000
250	PAR-38	Q250PAR38SP	Spot	26	34,000	4000
250	PAR-38	Q250PAR38FL	Flood	60	6000	4000
300	PAR-56	300PAR56/NSP	Spot	15X20	70,000	2000
300	PAR-56	300PAR56/MFL	Med-Flood	20X35	22,000	2000
300	PAR-56	300PAR56/WFL	Wide-Flood	30X60	10,000	2000
500	PAR-56	Q500PAR56/NSP	Spot	15X32	90,000	4000
500	PAR-56	Q500PAR56/MFL	Med-Flood	20X42	49,000	4000
500	PAR-56	Q500PAR56/WFL	Wide-Flood	34X66	18,000	4000
1000	PAR-64	Q1000PAR64NSP	Spot	14X31	160,000	4000
1000	PAR-64	Q1000PAR64MFL	Med-Flood	22X45	60,000	4000
1000	PAR-64	Q1000PAR64WFL	Wide-Flood	45X72	27,000	4000

UNDERWATER LIGHTING

Table 17-3. Arrangements of Lights Surrounding the Jet Nozzle in Fig. 17-3

No. Lamps	Color	Type	Candlepower (each lamp)
4[1]	blue	Q1000PAR64NSP	160,000
3[1]	red	Q1000PAR64NSP	160,000
3[1]	amber	Q500PAR56NSP	90,000
2[1]	clear	Q500PAR56NSP	90,000
Total 12			1,570,000
8[2]	blue	Q500PAR56MFL	49,000
8[2]	green	Q500PAR56MFL	49,000
8[2]	amber	Q300PAR56MFL	22,000
Total 24			960,000

[1] The outside ring is lighted by 12 fountain lights located on a 2½-foot radius.
[2] The outside ring is lighted by 24 fountain lights located on a 6-foot radius.

light. If this condition exists, it is recommended that clear, amber, or turquoise colors be used.

Heights to Be Illuminated

Table 17-1 may be used as a guide in determining the minimum beam candlepower required for a given height in order to achieve a reasonable balance of decorative effect colors. The values are based upon the use of standard lenses as manufactured by Kim Lighting & Manufacturing Co., Inc.

The total footcandle requirements for a specific lighting layout would be contingent upon the perimeter of the layout.

A tabulation of standard available lamps and their rated candlepower can be found in Table 17-2.

Typical Layout

The illustration in Fig. 17-3 shows a typical fountain layout. In the first part of Table 17-3, the

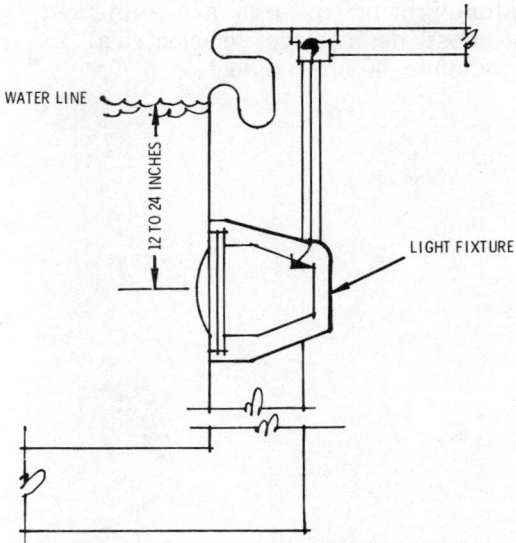

Fig. 17-4. A typical swimming pool layout.

twelve fountain lights surrounding the jet nozzle are located on a 2½-foot radius. In the second part of Table 17-3, the outside ring is lighted by 24 fountain lights located on a 6-foot radius.

Swimming Pool Lighting

There is a large selection of underwater lights, from small 75-watt units to large 1000-watt units, to meet practically every underwater lighting requirement.

In selecting equipment for underwater lighting, the greatest economy is achieved through the selection of a high quality fixture designed to give years of dependable service. By their very nature, un-

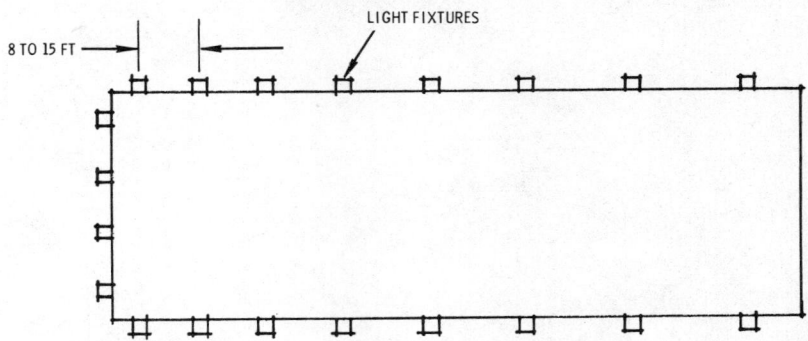

Fig. 17-5. Recommended lighting layouts for swimming pools.

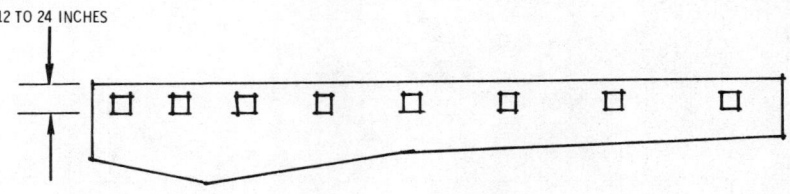

PRINCIPLES OF ILLUMINATION

derwater lighting fixtures are subjected to the forces most destructive to electrical fixtures—water, chemicals, and neglect.

The illustrations in Figs. 17-4 and 17-5 show some recommended lighting layouts for swimming pools.

UNIT 18

Roadway Lighting

Street and roadway lighting has become so common that many of us fail to notice or appreciate its full value. But when one learns of the benefits derived, in the reduction of traffic accidents, crime and vagrancy, and increased business in well-lighted shopping areas, we find that it is a very important branch of illuminating engineering.

Incandescent lamps of several hundred watts were once used extensively for street and roadway lighting. However, these are being rapidly replaced by high-pressure sodium lamps (as described in Unit 6) because the latter have greater efficiency and longer life.

Effective roadway and street lighting can only be achieved through careful and intelligent planning and by following the American Standard Practice for Street and Highway Lighting. The following should be taken into consideration in the order given:

1. The area and roadway classification.
2. The proper illumination level for the classification.
3. Selection of lighting fixtures according to the requirements.
4. The proper location of the lighting fixtures to provide the required quantity and quality of illumination.

FOOTCANDLES ACCORDING TO ROADWAY CLASSIFICATION

The average horizontal footcandles recommended in Table 18-1 represent average illumination on the roadway pavement when the light source is at its lowest output and also when it is in its dirtiest condition.

In lighting urban roadways, the proper classification of areas and streets is important. The selection of the proper type of light source and its application is also important to a good lighting job.

It should be remembered that the illumination values given in Table 18-1 are minimum maintained in service. Also, for commercial reasons, much higher levels are generally used in the commercial district than those minimum levels required for traffic safety.

Table 18-1. Average Illumination on Roadway Pavement (Light at Lowest Output)

	AREA CLASSIFICATION		
Roadway Classification	Downtown (footcandles)	Intermediate (footcandles)	Outlying Urban and Rural (footcandles)
Major	2.0	1.2	0.9
Collector	1.2	0.9	0.6
Local or Minor	0.9	0.6	0.2
EXPRESSWAYS AND FREEWAYS			
Roadway Classification	Expressways (footcandles)	Freeways (footcandles)	
Continuous Urban	1.4 to 2	0.6	
Continuous Rural	1.0	0.6	
Interchange Urban	2.0	0.6	
Interchange Rural	1.4	0.6	

TYPICAL APPLICATIONS

The following are typical applications furnished by Westinghouse Electric Corporation and are in accordance with the ASA-IES Standards.

MAJOR STREETS

Downtown

Fig. 18-1 shows a typical layout for a 70-foot street using 1000-watt mercury lamps in a NEMA

PRINCIPLES OF ILLUMINATION

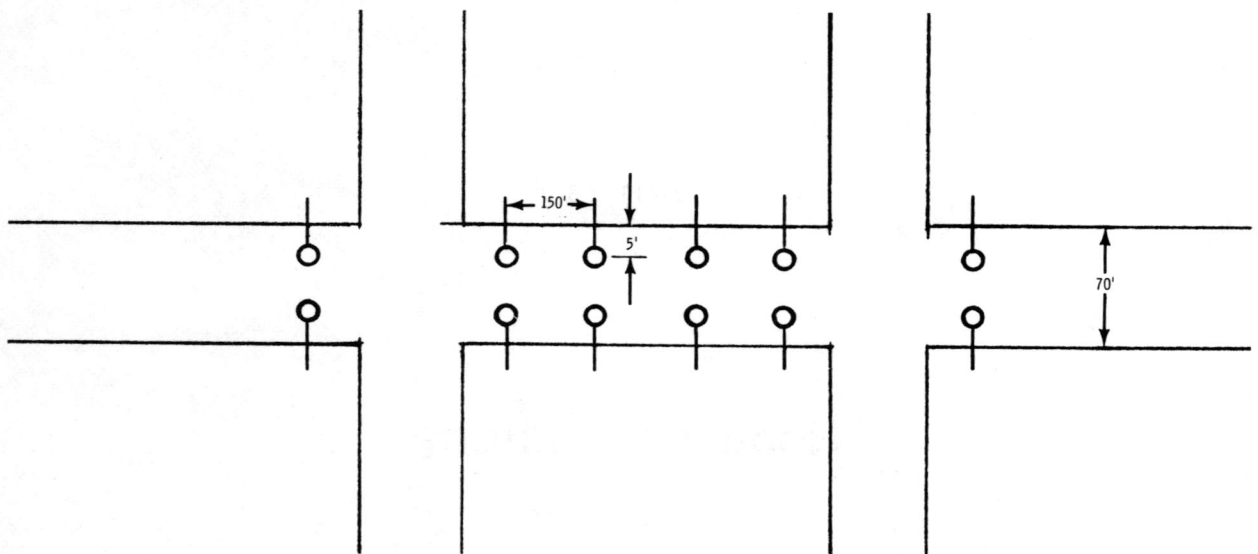

Fig. 18-1. A typical lighting layout for a 70-foot street using 1000-watt mercury lamps.

Type III lighting fixture with a short spacing ratio and a semi-cutoff. They are mounted at 33 feet in an opposite arrangement and spaced 150 feet on centers for a calculated average of 2.7 footcandles. This meets ASA-IES recommendations.

Fig. 18-2 illustrates a typical plan for the same street but with a higher level of illumination. Two 1000-watt mercury lighting fixtures are used on each pole, either side by side or one above the other. This arrangement gives a calculated average of 5.3 maintained footcandles.

If a higher level of illumination is required, high pressure sodium lamps may be used instead of mercury.

Intermediate Area

Fig. 18-3 illustrates a layout to provide for either an intermediate area or collector street in a downtown area. Here, the 400-watt clear mercury lamp is used in a type III medium spacing ratio, semi-cutoff luminaire, or a type IV inside frosted phosphor coated lamp is used. Mounting height is 30 feet. The arrangement provides an average of 1 to 2 maintained footcandles.

Outlying Area

The layout in Fig. 18-4 is for an outlying area or collector street in an intermediate area. It uses a

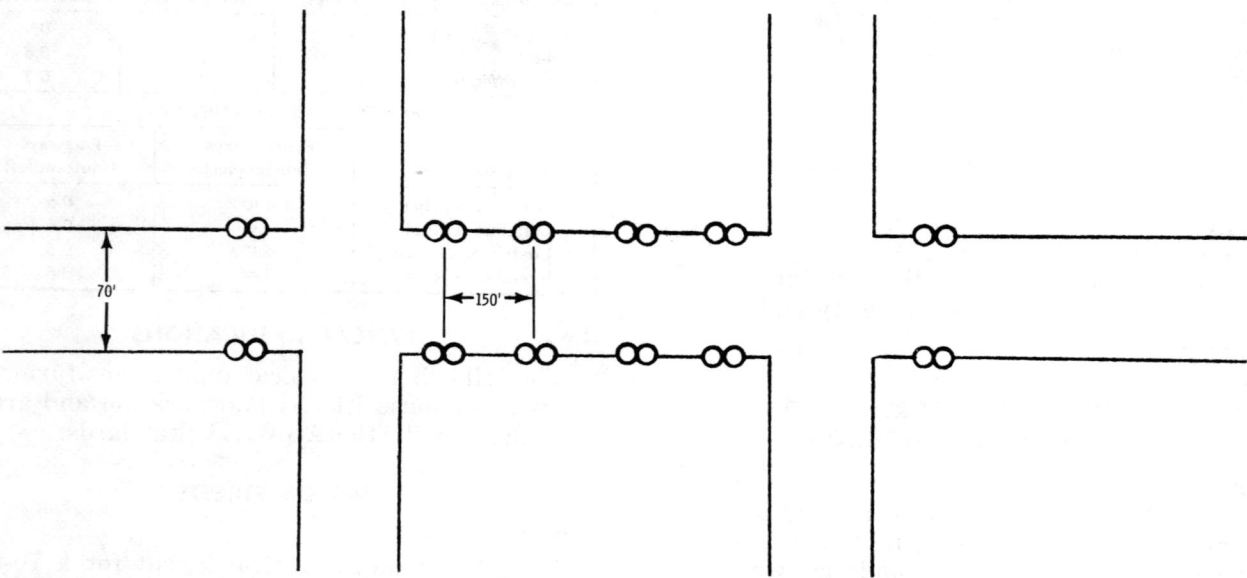

Fig. 18-2. A lighting layout for a 70-foot street with a high level of illumination requirement.

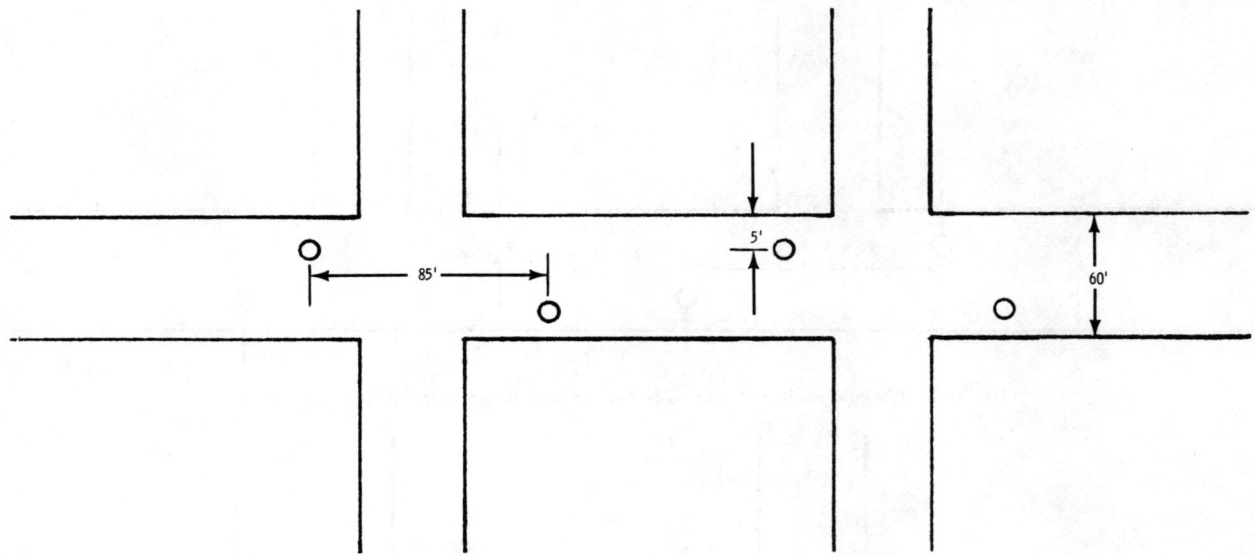

Fig. 18-3. A typical lighting layout for an intermediate area or collector street in a downtown area.

400-watt mercury lamp in a type III, medium spacing ratio, and semi-cutoff. The mounting height is 30 feet and provides an average of 0.9 maintained footcandles.

COLLECTOR STREET

The layout in Fig. 18-5 is for a collector street in an outlying area or a local street in an intermediate area. It uses a 175-watt mercury lamp in a type II lighting fixture, medium spacing ratio, and semi-cutoff with a mounting height of 25 feet for an average of 0.6 maintained footcandles.

LOCAL AND MINOR STREET

The layout in Fig. 18-6 uses a 175-watt mercury lamp in a type II, long spacing ratio, and semi-cutoff luminaire at 25 feet mounting height for an average of 0.35 maintained footcandles.

RESIDENTIAL-OUTLYING AREA

The layout in Fig. 18-7 uses a 100-watt mercury lamp in a type II, long spacing ratio, and semi-cutoff luminaire with a mounting height of 25 feet and an average of 0.2 maintained footcandles.

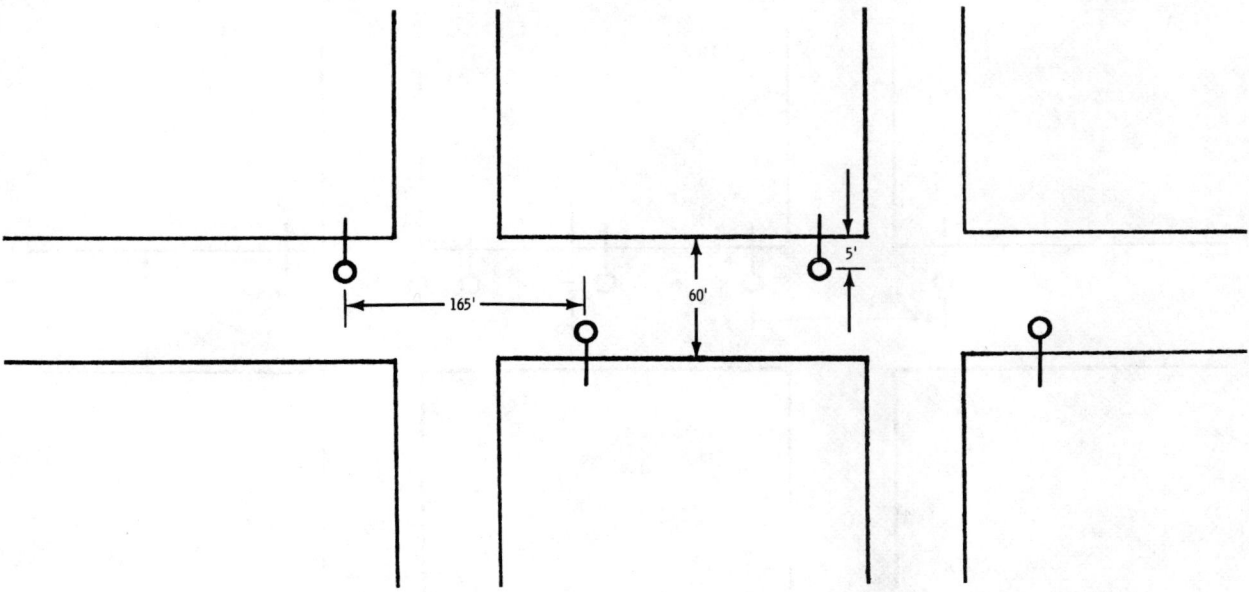

Fig. 18-4. This lighting layout uses 400-watt mercury lamps with medium spacing.

PRINCIPLES OF ILLUMINATION

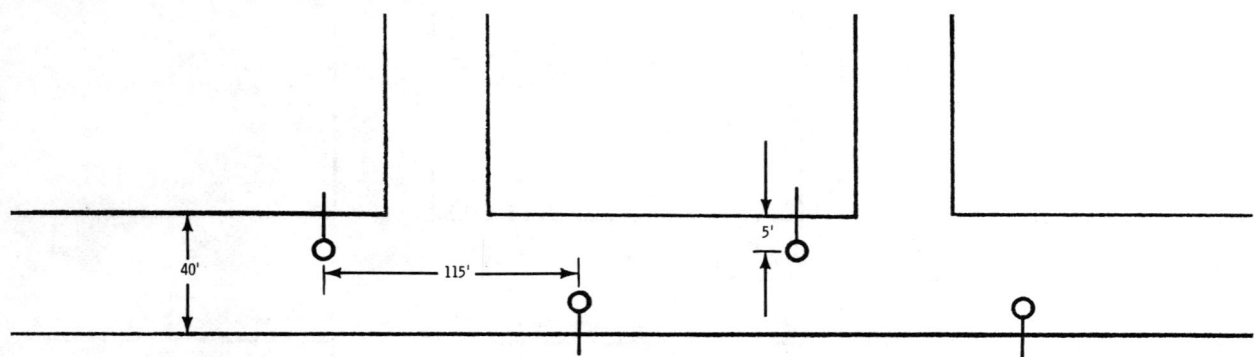

Fig. 18-5. This lighting layout uses 175-watt mercury lamps with medium spacing.

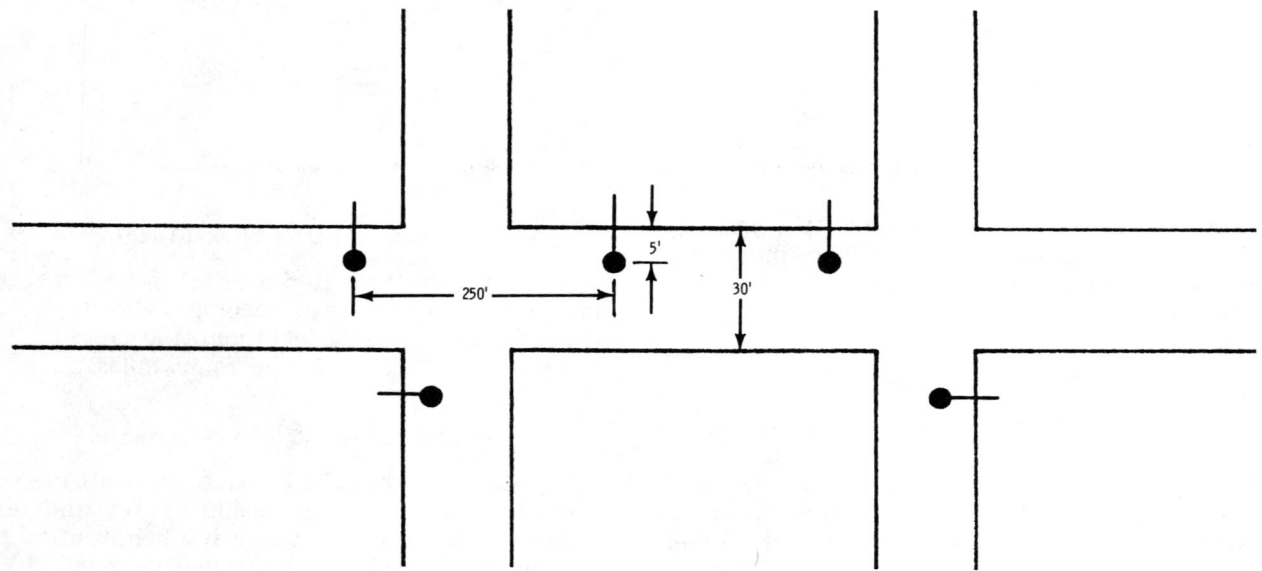

Fig. 18-6. This lighting layout uses 175-watt mercury lamps with long spacing.

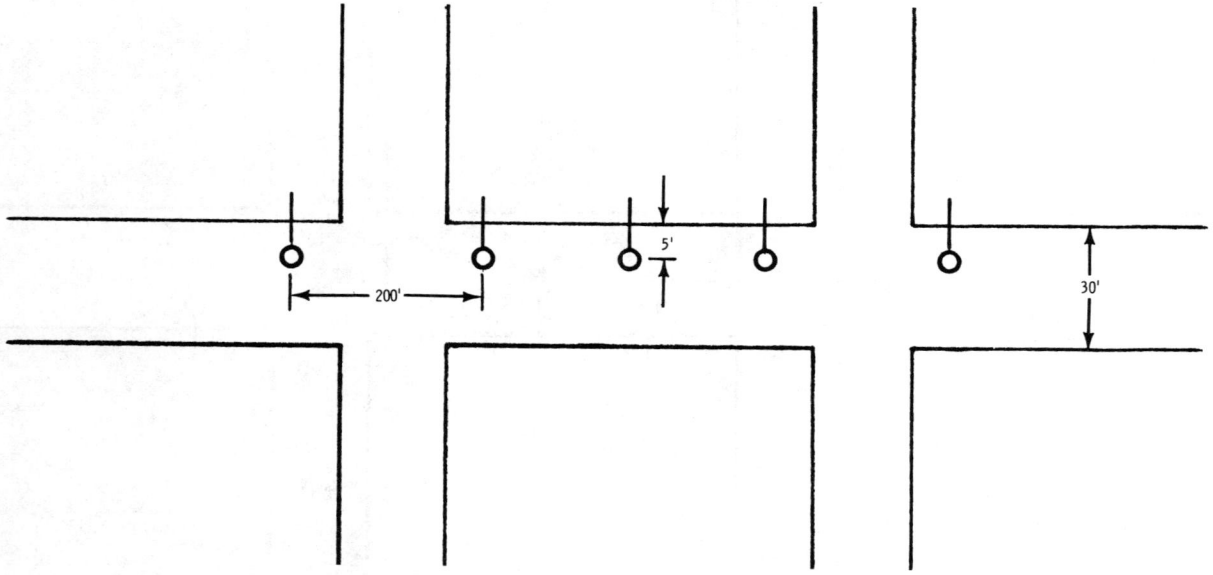

Fig. 18-7. This lighting layout uses 100-watt mercury lamps with long spacing.

114

ROADWAY LIGHTING

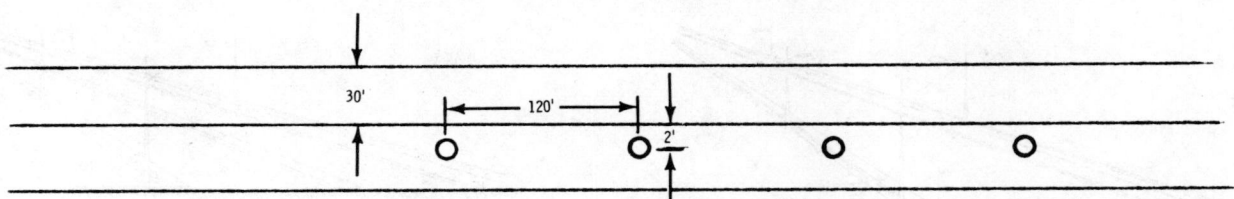

Fig. 18-8. This lighting arrangement of a residential area uses 175-watt mercury lamps.

The alternate lighting arrangement of the residential area in Fig. 18-8 uses a 175-watt mercury lamp in a type III pole-top mounted luminaire 13 feet above grade for an average of 0.25 maintained footcandles.

As a quick check reference guide to various application conditions, the charts in Fig. 18-9 may be used. These charts are based on the proper selection of a lighting fixture for the desired application conditions.

HIGHWAY, FREEWAY, AND EXPRESSWAY APPLICATIONS

Lighting fixtures on expressways or freeways should be located at a sufficient distance from the edge of the curb or traffic lane to minimize the chances of a serious accident should a car be forced from the regular traffic lane. If they are less than 30 feet from the edge of the curb or traffic lane, the poles should be protected by guard rails.

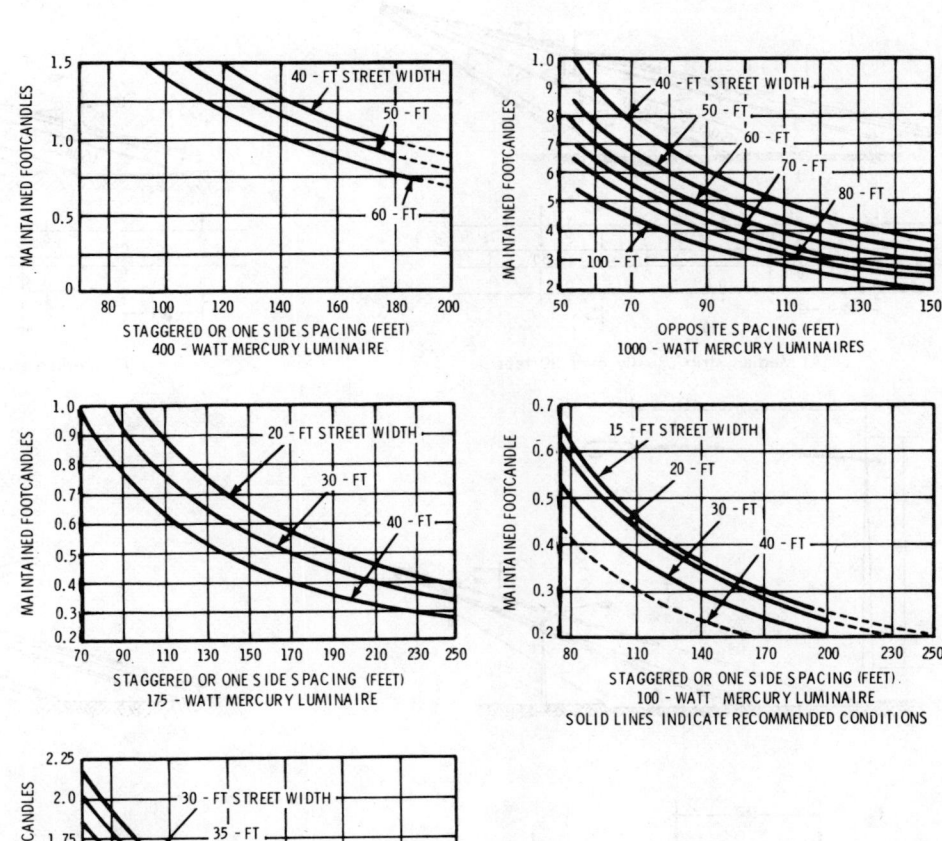

Fig. 18-9. A quick check reference guide to various applications.

PRINCIPLES OF ILLUMINATION

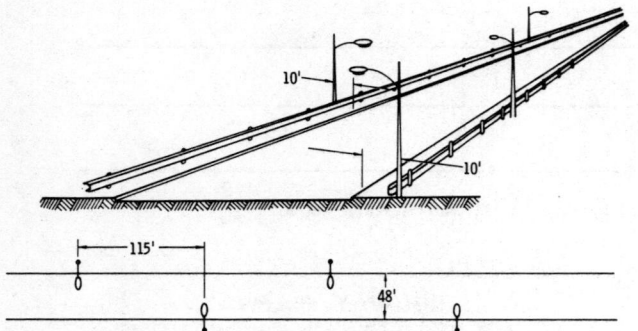

Fig. 18-10. A typical lighting arrangement of a roadway.

Where there is no median strip, pole arrangements should be located on the outside in either a staggered or opposite arrangement, depending upon the width of the roadway as shown in Fig. 18-10.

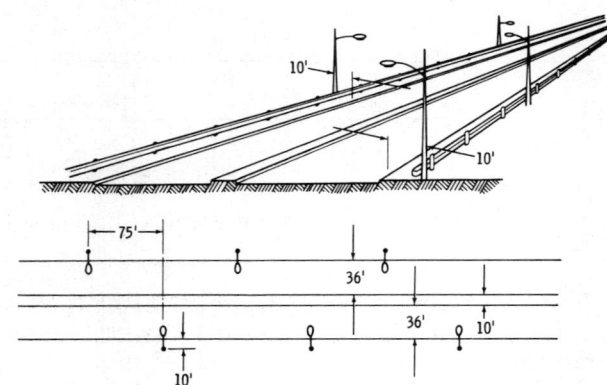

Fig. 18-11. A lighting arrangement of a divided roadway with a narrow median strip.

If the roadway is provided with a narrow median strip, and the total roadway width is not greater than 125 feet, then the entire area, includ-

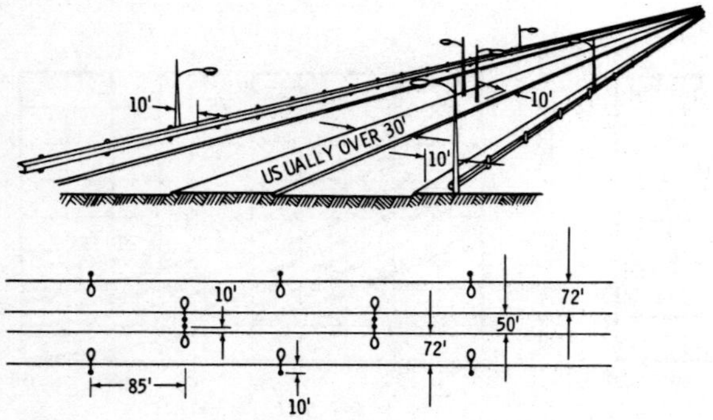

(A) Median strip usually over 30 feet.

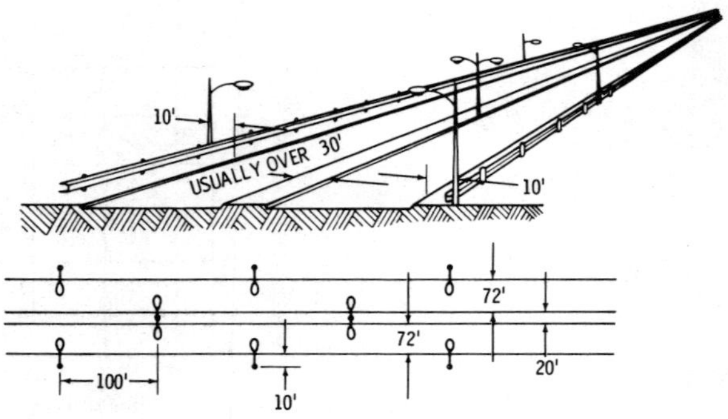

(B) Median strip usually under 30 feet.

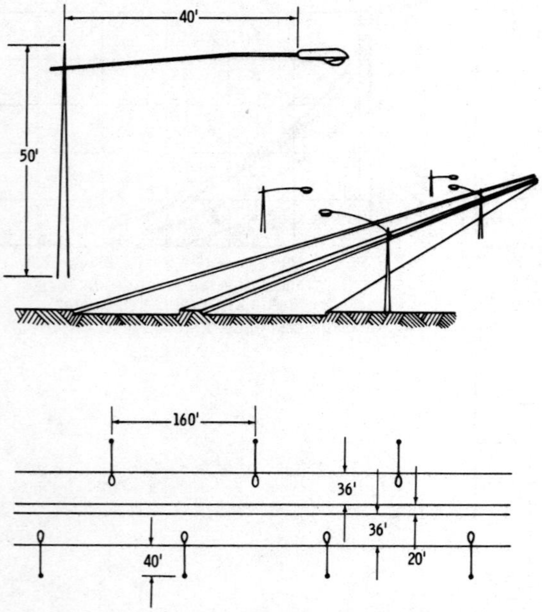

(C) Another method of roadway lighting.

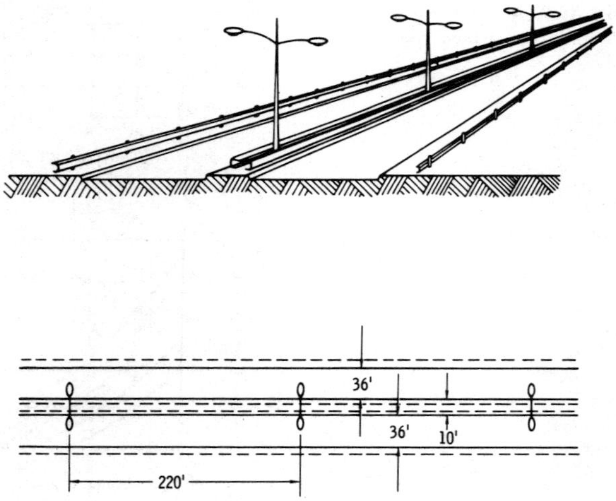

(D) Another method of roadway lighting.

Fig. 18-12. Various ways to illuminate roadways.

116

Table 18-2. Acceptable Roadway Lighting Configurations*

SIX-LANE EXPRESSWAYS
Road Type .... Six-lane, urban expressway, 10-foot median, 10-foot shoulders
Road Width .... 82 feet (six 12-foot lanes plus 10-foot median)
Desired Illumination Level .... 1.4 average horizontal footcandles
Desired Uniformity Ratio 3:1

	Lamp Data			Pole Data						Illumination[1] (footcandles)		
Type	ASA Designation	Light Output (lumens)	Power (watts)	Mounting Height (ft)	Overhang (ft)	Luminaire Type	Light-Distribution Type	Arrangement	Spacing (ft)	Average	Minimum	Uniformity Ratio
Mercury	H33-1CD	21,000	400	35	4	OV-25	III	Staggered	55	1.91	0.69	2.8:1
Mercury	H33-1CD	21,000	400	35	4	OV-25	III	Center-Opposite	136	1.40	0.57	1.9:1
Mercury	H35-18NA	39,000	700	40	4	OV-50	III	Opposite	200	1.69	0.73	2.3:1
Mercury	H35-18NA	39,000	700	40	4	OV-50	III	Staggered	121	1.40	0.64	2.2:1
Mercury	H36-15GV	55,000	1000	40	4	OV-50	III	Staggered	158	1.40	0.61	2.3:1
Mercury	H36-15GV	55,000	1000	50	4	OV-50	III	Staggered	162	1.40	0.45	3.1:1
Mercury	H36-15GV	55,000	1000	50	4	OV-50	III	Center-Opposite	220	1.5	0.60	2.5:1

[1] Based on 70% maintenance factor.

TWO-LANE ROADS
Road Type .... Two-line, major rural, 10-foot shoulders
Road Width .... 24 feet (two 12-foot lanes)
Desired Illumination Level .... 0.9 average horizontal footcandles
Desired Uniformity Ratio 3:1

	Lamp Data			Pole Data						Illumination[1] (footcandles)		
Type	Designation	Light Output (lumens)	Power (watts)	Mounting Height (ft)	Overhang (ft)	Luminaire Type	Light-Distribution Type	Arrangement	Spacing (ft)	Average	Minimum	Uniformity Ratio
Mercury	H39-22KB	7,700	175	30	4	OV-15	II	One Side	92	0.9	0.36	2.5:1
Mercury	H37-5KB	12,100	250	30	4	OV-15	II	One Side	135	0.9	0.49	1.8:1
Mercury	H33-1CD	21,000	400	30	4	OV-25	II	One Side	100	2.13	0.71	3.0:1
Mercury	H33-1CD	21,000	400	30	4	OV-25	II	Staggered	150	1.41	0.47	3.0:1
Mercury	H35-18NA	39,000	700	35	4	OV-50	III	One Sde	200	1.49	0.51	2.9:1
Mercury	H35-18NA	39,000	700	35	4	OV-50	III	Staggered	210	1.47	0.49	3.0:1

[1] Based on 70% maintenance factor.

FOUR-LANE EXPRESSWAYS
Road Type .... Four-lane, rural expressway, 36-foot median, 10-foot shoulders
Road Width .... 84 feet (four 12-foot lanes plus 36-feet median)
Desired Illumination Level .... 0.6 average horizontal footcandles
Desired Uniformity Ratio 3:1

	Lamp Data			Pole Data						Illumination[1] (footcandles)		
Type	ASA Designation	Light Output (lumens)	Power (watts)	Mounting Height (ft)	Overhang (ft)	Luminaire Type	Light-Distribution Type	Arrangement	Spacing (ft)	Average	Minimum	Uniformity Ratio
Mercury	H33-1CD	21,000	400	30	4	OV-25	III	Staggered	50	2.04	0.76	2.7:1
Mercury	H33-1CD	21,000	400	30	4	OV-25	III	Opposite	100	2.04	0.75	2.7:1
Mercury	H33-1CD	21,000	400	35	4	OV-25	III	Staggered	60	1.67	0.60	2.8:1
Mercury	H35-18NA	39,000	700	35	4	OV-50	III	Staggered	90	1.95	0.67	2.9:1
Mercury	H35-18NA	39,000	700	40	4	OV-50	III	Staggered	115	1.45	0.58	2.5:1
Mercury	H35-18NA	39,000	700	40	4	OV-50	III	Center-Opposite	210	1.66	0.57	2.9:1
Mercury	H35-18NA	39,000	700	40	4	OV-50	III	Opposite	200	1.40	0.48	2.9:1
Mercury	H36-15GV	55,000	1000	50	4	OV-50	III	Staggered	210	1.6	0.50	3.0:1
Mercury	H36-15GV	55,000	1000	50	4	OV-50	III	Staggered	240	1.4	0.46	3.0:1
Mercury	H36-15GV	55,000	1000	50	4	OV-50	III	Center-Opposite	210	1.5	0.60	2.5:1

[1] Based on 70% maintenance factor.
*Courtesy Westinghouse Electric Corp.

PRINCIPLES OF ILLUMINATION

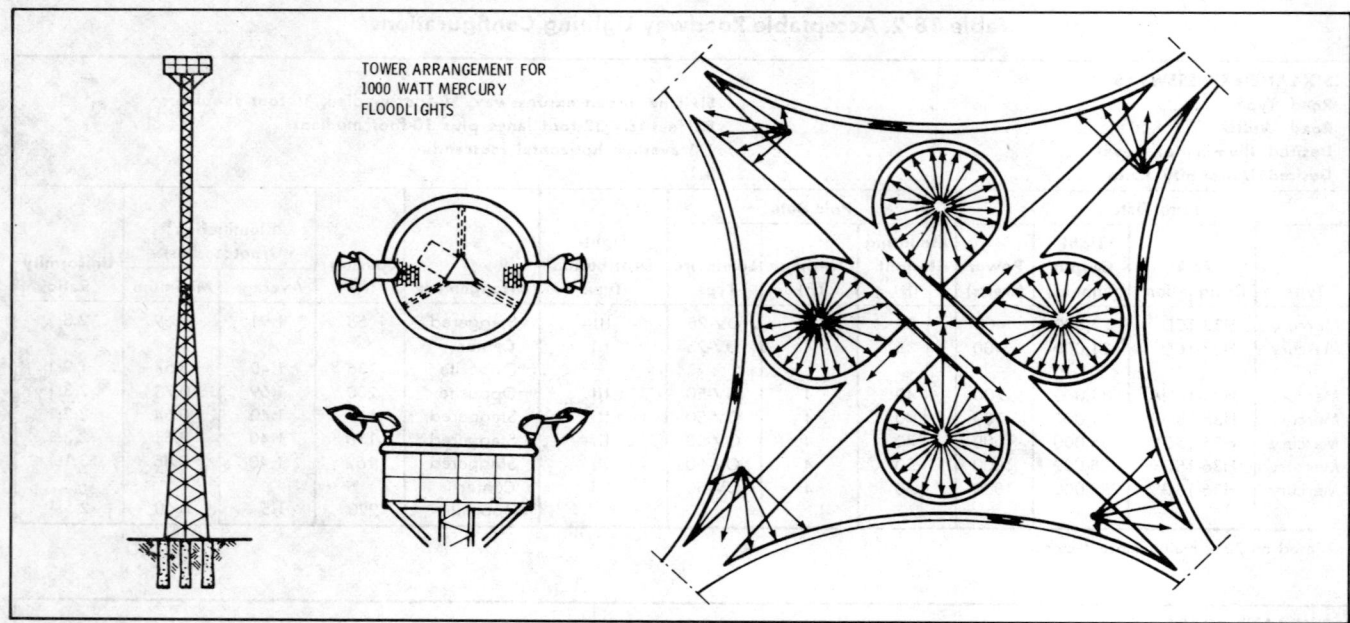

Fig. 18-13. Lighting interchange by using tower.

ing the median, should be treated as a single roadway as illustrated in Fig. 18-11.

If the median strip exceeds 30 feet in width, and the two roadways, including the median strip, exceed 125 feet, then the poles (standards) should be mounted in the median strip, and each roadway should be treated as a separate road as shown in Fig. 18-12A.

If the median strip is less than 30 feet in width, the pole locations may be arranged as shown in Fig. 18-12B.

Figs. 18-12C and 18-12D show two other suitable methods for mounting poles and fixtures.

Table 18-2 gives acceptable roadway lighting configurations for six-lane expressways, two-lane roads, and four-lane expressways.

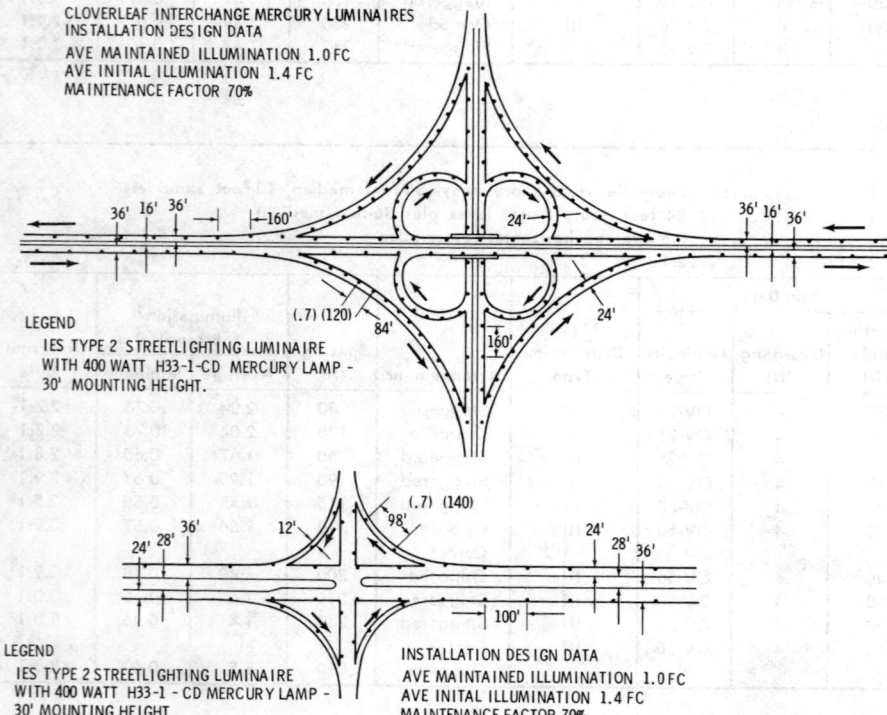

Fig. 18-14. Typical layout for cloverleaf interchange and diamond intersection.

118

HIGHWAY INTERCHANGE LIGHTING

At a highway interchange, the lighting system should be so arranged that it will aid the driver in selecting the roadway to his destination, minimizing his chances of confusion and making such roadways more comfortable and convenient for night driving (Fig. 18-13).

These exit and entrance roadways are usually more narrow than the horizontal roadways they connect. They are most often constructed on a horizontal curve, but may also be on a grade curve, making the placement of lighting fixtures difficult for the maximum utilization.

The specific lighting level for interchanges should be maintained as a minimum. Preferably, it should be a little higher than that on any of the adjacent roadways.

To light interchanges, on otherwise unlighted roadways, it is desirable to start the lighting of the main through road approximately ½ mile before the first exit road and graduate the intensities from about 0.3 footcandle for the first ¼ mile, to 0.7 footcandle for the second ¼ mile. Then use from 1 to 1¼ footcandles through the interchange, and gradually taper off in this reverse order. All values are minimum maintained.

The illustrations shown in Fig. 18-14 provide some typical layouts for several different types of interchanges.

UNDERPASS LIGHTING

Medium-length underpasses, 75 to 150 feet, usually require nightime lighting only. Generally, underpasses in excess of 150 feet will require lighting both day and night.

The horizontal illumination on the roadway for nightime lighting should be at least twice that of the adjacent roadway illumination, and preferably more. Good recommended practice would be 5 to 10 footcandles.

Underpasses often do not have solid walls finished with high reflectance surfaces. The ceiling or overhead structure may be broken up by cross beams which make it impossible to create uniform brightness of the ceiling area. In this case, regardless of the lamp type that is chosen, the lighting fixtures should be equipped with some form of optical control that will direct the light to the roadway.

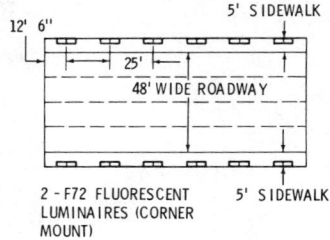

(A) Fluorescent system.

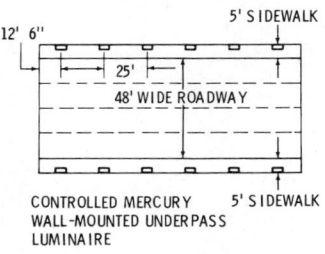

(B) Mercury system.

Fig. 18-15. Some typical underpass lighting layouts.

A typical layout for fluorescent and mercury lighting, each providing approximately the same level of illumination, is shown in Fig. 18-15.

Underpass lighting fixtures may be either the fluorescent, high-pressure sodium, or mercury vapor lamp types.

SUMMARY

- The major purposes of roadway lighting are:
  1. To provide for traffic safety.
  2. To curb crime and vagrancy.
  3. To promote civic progress.
- Effective roadway lighting must be well planned, taking into account the following considerations, in order:
  1. The area and roadway classification.
  2. The proper illumination level for the classification.
  3. Selection of fixtures according to the light distribution required.
  4. The proper role and fixture locations (mounting height, overhang, and spacing) to provide the required quantity and quality of illumination.

UNIT 19

Sign Lighting

Electric signs have been used for advertising, identification of buildings, and for displays almost since the invention of the electric lamp. Highway information as to routes, cities, etc., also makes use of illuminated signs in order to help make night viewing easier for the driver. Electric sign types in use today include:

1. Exposed incandescent lamp signs.
2. Exposed luminous tube signs.
3. Internally lighted luminous signs.
4. Silhouette signs.
5. Floodlighted signs.

The functions of an electric sign are:

1. To attract attention.
2. To create a favorable impression.
3. To characterize the type of business being advertised.
4. To convey information.

The last three functions are usually dependent on the artistic ability of the sign designer. However, the first function, the attention-attracting value of the sign, is highly influenced by the method of lighting, since all of us instinctively glance at high-brightness areas. In general, the higher the brightness, the greater the attracting value.

BILLBOARD LIGHTING

One of the simplest forms of illuminated signs is the billboard type as illustrated in Fig. 19-1. Billboards of this type are quite commonly equipped with electric lights because, in many cases, they actually attract more attention when lighted at night than they do during daylight hours.

Early billboards were almost always lighted with simple porcelain-enamel incandescent floodlights. Today a wide variety of incandescent and fluorescent equipment is available, which gives the lighting designer much more freedom in tailoring the lighting to fit the needs of all signs. Mercury lamps are now beginning to find their way into the sign lighting market, especially since the improved color lamps have become available.

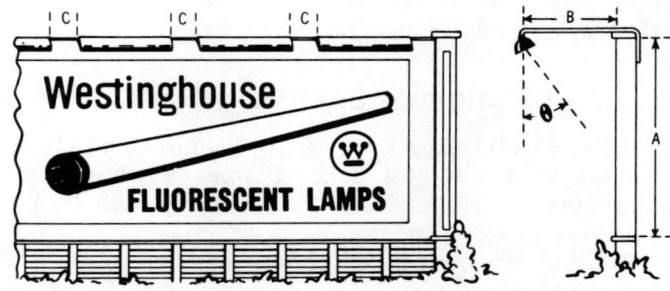

Fig. 19-1. Simple form of billboard illumination.

Table 19-1. Illumination Level With Incandescent Lamps

Dimensions (ft) (Fig. 19-1)			Aiming Angle θ (degrees)	Footcandles			
A	B	C		400-Watt Metal Halide Lamp	500-Watt Tungsten-Halogen Lamp	1500-Watt Tungsten-Halogen Lamp	400-Watt Mercury
12	4	8	40	55	—	115	—
12	4	6	40	—	50	—	55
15	4¾	7⅔	40	50	—	100	—
15	4¾	5¾	40	—	55	—	50

PRINCIPLES OF ILLUMINATION

Table 19-2. Illumination Levels With Fluorescent Lamps

Dimensions (ft) (Fig. 19-1)			Aiming Angle θ (degrees)	Footcandles			
				Open Unit		Enclosed Unit	
				Super-Hi Lamp 1500 mA	High-Output Lamp 800 mA	Super-Hi Lamp 1500 mA	High-Output Lamp 800 mA
A	B	C					
4	2	2	40	105	75	100	75
6	3	3	40	70	50	70	50
8	4	3	40	37	26	35	25
10	4	3	30	25	18	24	17
12	4	3	30	21	15	20	15
14	6	3	30	18	12	17	12
18	8	3	35	14	10	13	10

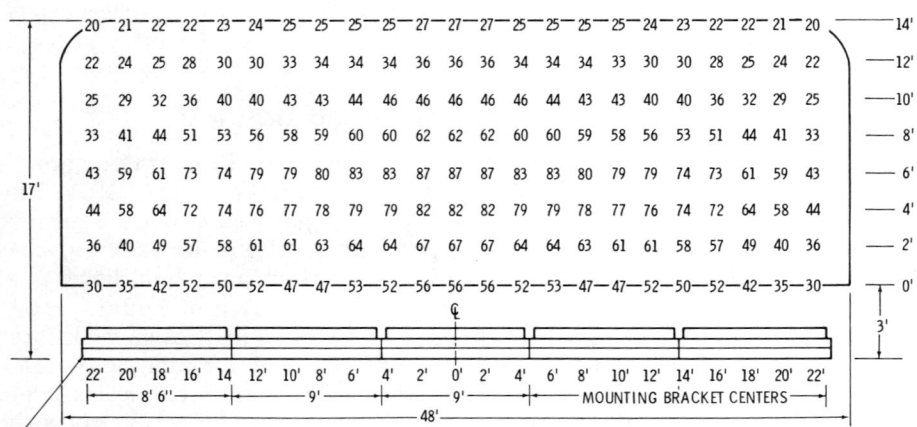

Fig. 19-2. Using bottom-mounted floodlights to illuminate billboard.

The illumination and the location of the lighting equipment are of primary concern in the design of sign lighting. Levels of illumination obtained with incandescent lamps are shown in Table 19-1, and those obtained with fluorescent lamps are shown in Table 19-2.

LOCATION OF LIGHTING FIXTURES

Billboard lighting fixtures are usually located at the top or bottom of a sign. At times, when a very high level of illumination is required, both locations are used.

Bottom-mounted floodlights are easier to maintain, and they avoid the problems of daytime shadows and nighttime reflected glare—this principle is illustrated in Fig. 19-2. They are, however, more subject to vandalism, and the lighting equipment should be completely enclosed and weathertight.

Top-mounted lighting fixtures, at times, may be more convenient to install, but this mounting type has the disadvantage of casting shadows on the sign copy in daytime, and at night they produce light-source reflections on shiny copy surfaces. In some cases, the reflected glare from top-mounted fixtures may project at just the exact angle to strike the eyes of observers who are slightly below the board. This glare is not only uncomfortable to observers but can be dangerous in the case where observers may also be drivers of motor vehicles.

Regardless of whether the lighting source is top or bottom mounted, the lamps must be far enough

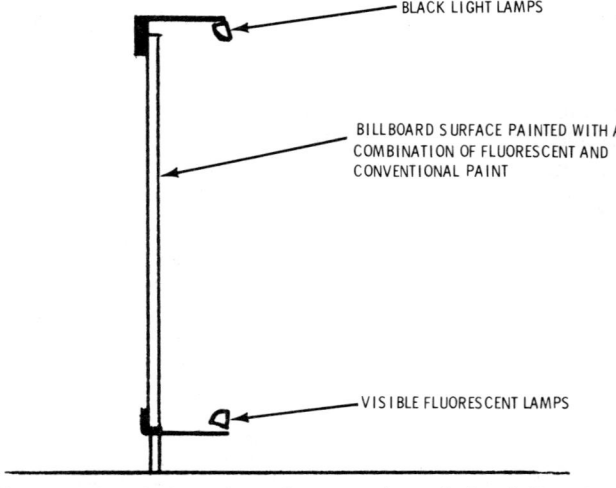

Fig. 19-3. A typical two-lamp fluorescent lamp flasher ballast circuit.

out in front of the sign to produce reasonably uniform illumination. This placement of the lamps will also help to reduce shadows and glare on the sign being illuminated.

BLACK-LIGHT USES

Black light, as discussed in Section II, may be used in sign lighting with reasonable effectiveness. Signs painted with fluorescent paint and exposed to black light give the illusion of glowing without any apparent light source being applied, since the ultraviolet radiation of black light is invisible to our eyes.

Sometimes, black light can be effectively utilized with visible light for illuminating a sign as shown in Fig. 19-3. Continued fluorescent lamps are combined with black-light lamps to illuminate the surface of the billboard. This surface contains copy which is visible in daylight hours and under the fluorescent lamps. In addition, there is other copy applied to the surface in fluorescent paint. The colors and the pattern of this fluorescent copy are not visible during the daylight hours or under the white light from the fluorescent lamps. However, under the ultraviolet radiation of the black-light lamps, the fluorescent copy appears in brilliant colors.

FLASHING

A basic type of flashing light is one that flashes off-on in rapid succession. It is one of the finest eye-catchers in sign lighting, since it combines the two prime factors of attention—brightness and motion.

A simple form of flasher can be found built into an incandescent Christmas tree lamp. Such a lamp will flash automatically as long as it is energized, since the heat from the filament heats a movable bimetallic strip which bends away from the lead-in wire causing the lamp to go out (breaks the circuits). As the strip begins to cool, it gradually bends back toward the lead-in wire. When the circuit is closed again, the lamp comes on. This process is repeated as long as the lamp is energized, and it causes a continuous flashing of the lamp.

Rapid-start fluorescent lamps can also be flashed without damage to the electrodes, so long as these electrodes are kept heated to the correct temperature at all times by using a special flasher ballast. A typical two-lamp fluorescent lamp flasher ballast circuit is shown in Fig. 19-4.

The lighting effect illustrated in Fig. 19-3 can be made even more effective by flashing the convenient fluorescent lamps alternately with the black-light lamps.

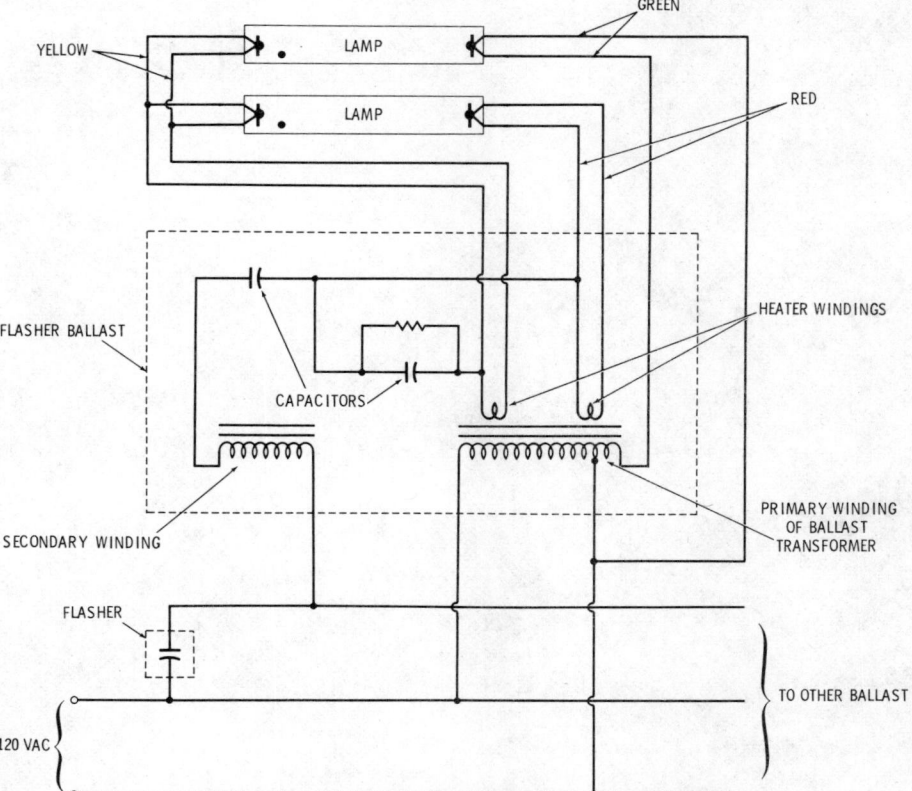

Fig. 19-4. Using black-light lamps in combination with fluorescent lamps for effective billboard lighting.

EXAMPLE OF BILLBOARD LIGHTING DESIGN

Assume a billboard with a dark surface is located in dark surroundings as illustrated in Fig. 19-5. Design a suitable lighting system for the illumination of this billboard, using fluorescent lamps.

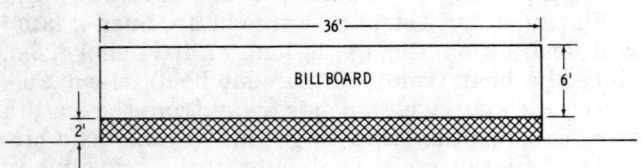

Fig. 19-5. Billboard with a dark surface used to design a suitable lighting system.

First, determine the recommended level of illumination required. Appendix A gives the recommended level of illumination as 50 footcandles for bulletins and poster boards in dark surroundings with a dark surface.

Second, refer to the proper table to determine the required equipment. Table 19-2 gives 50 footcandles of illumination with 800-mA high-output lamps for a 6-foot high surface, aimed at 40°, mounted 3 feet out from the sign, and spaced 3 feet apart.

Third, begin to lay out the system.

SUMMARY

- Electric signs are used for advertising, displays, identification, and conveying of information.

- Bottom-mounted floodlights are easier to maintain, and they overcome the problems of daytime shadows and nighttime reflected glare.

- Top-mounted floodlights are sometimes more convenient to install and are less subject to vandalism.

- A combination of white light and black light on fluorescent paint can produce very dramatic and interesting effects in billboard lighting.

SECTION VI

SPECIAL LIGHTING APPLICATIONS

UNIT 20

Germicidal Lamps

Germicidal lamps are sources of radiant energy at wavelengths just shorter than those of visible sunlight. These waves are destructive to bacteria and mold spores.

The source of ultraviolet energy in a germicidal lamp is a low-pressure arc operating on the same basic principle as the arc in a conventional fluorescent lamp. However, in germicidal lamps, there are no fluorescent powders or phosphors within the glass envelope or tube, and this tube is made of special clear glass which transmits the ultraviolet generated by the arc and directly released for air irradiation and the killing of air-borne bacteria and mold spores. It also aids in the deodorization of surrounding air.

LAMP TYPES

Germicidal lamps are manufactured in two types: the cold cathode and the slimline.

The majority of germicidal lamps operate most efficiently in still air at room temperature and when ultraviolet output is measured at an ambient temperature of 77°F. Temperature either higher or lower than this will cause the lamp to decrease in its output.

Slimline lamps are designed exceptions to this rule. Cooling of the bulb does not have the same adverse effects as with other lamps and for this reason, the slimline lamp is suited for use in air-conditioning ducts for the purification of the air.

Since germicidal lamps can produce sunburn, the American Medical Association has set a limit of 0.1 microwatt per square centimeter for continuous exposure and 0.5 microwatt for 7 hours per day. The latter may be used for classroom application or similar application where persons will not be exposed more than 7 hours per day.

APPLICATIONS

The lighting fixture, as illustrated in Fig. 20-1, is designed for germicidal lamps and is also designed to confine the radiation above the heads of the occupants in the room where used. Louvered equipment should be used where ceilings are less than nine feet high in order to avoid localized high concentration of ultraviolet flux which may be reflected down onto occupants.

Hospitals, schools, and offices are the most usual applications for germicidal lamps. The upper air disinfection method is effective at the breathing level of room occupants and will maintain a freedom from respiratory disease organisms comparable with outdoor air. Personal movement, body

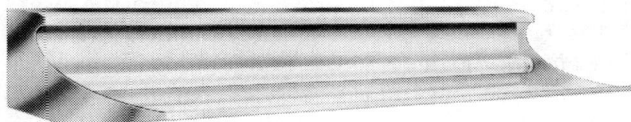

Courtesy Edwin F. Guth Co.

Fig. 20-1. A germicidal lamp designed to confine the radiation above the heads of occupants.

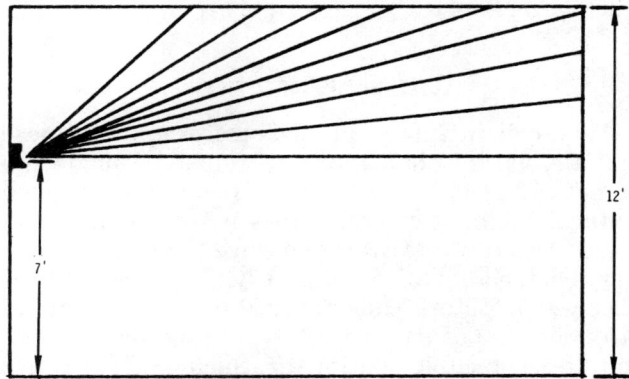

Fig. 20-2. Open germicidal light fixture used in room with ceiling height of nine feet or more.

PRINCIPLES OF ILLUMINATION

heat, and winter heating methods create convection currents through the entire room which are sufficient to equal 1 to 2 changes of air per minute.

An open germicidal light fixture used in a room with a ceiling height of nine feet or more is shown in Fig. 20-2.

See Fig. 20-3 for an illustration of louvered fixtures used where ceilings are less than nine feet.

AIR-DUCT INSTALLATION

It is possible to provide a sufficiently high level of ultraviolet in air ducts at usual air velocities. Tests have shown 90 to 99% of most bacteria was killed in a very short exposure time. The limitation of this method is that it can only make the duct air equivalent to good outdoor air, and its value is in the treatment of recirculated air and contaminated outdoor air.

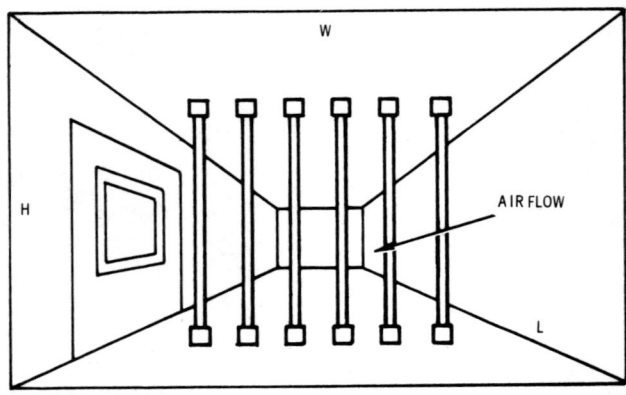

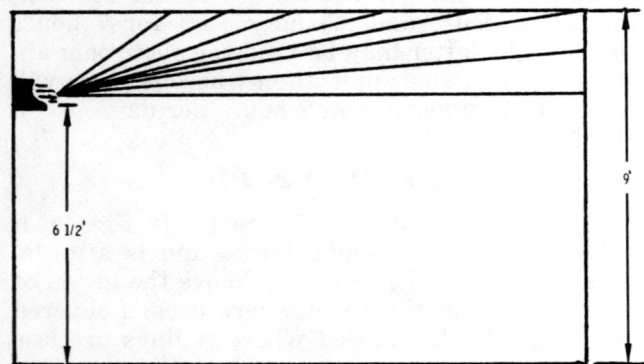

Fig. 20-3. Louvered fixtures used where ceiling height is less than nine feet.

Where possible, the best location of the lamps is across the duct. This will secure longer travel of the energy before absorption on the duct walls and promote turbulence to offset the variation in ultraviolet level throughout the irradiated part of the duct (Fig. 20-4).

WATER DISINFECTION

Water disinfection methods are similar to those for air, except that allowance must be made for some ultraviolet absorption by traces of such natural chemical contaminants as iron compounds. An 8 to 10 times greater exposure for wet than for dry bacteria is also necessary. The product of these two factors calls for water disinfection ultraviolet exposures of 700- to 1000-microwatt minutes per cubic centimeter. Such exposures are secured by slow gravity flow of water through shallow tanks under banks of lamps or by lamps in

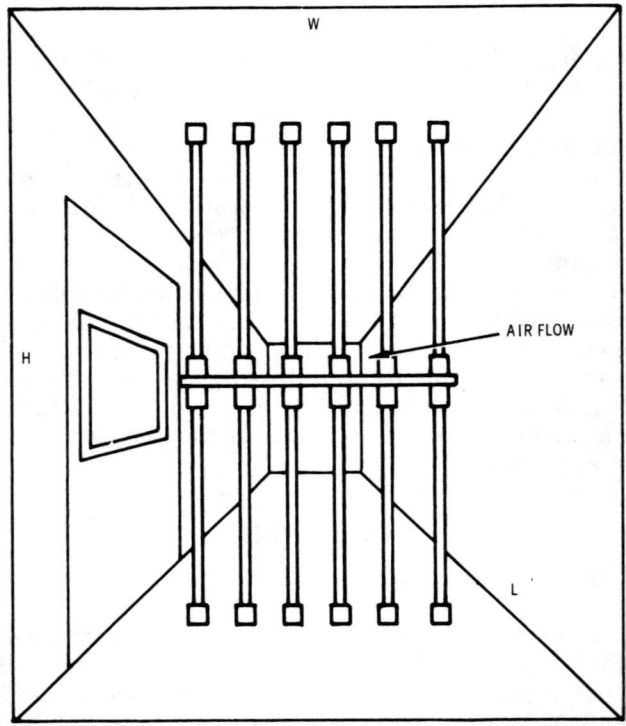

Fig. 20-4. Lamps mounted in air ducts.

isolating jackets of ultraviolet transmitting glass, immersed directly in the water.

Westinghouse Electric has been developing a packaged water purifier, using their germicidal lamps, for use in disinfecting swimming pool water. This device should be very useful for residential and private pools, especially where the users of the pools may be allergic to chlorine. Information on this device should be available from Westinghouse at the time of publication of this book. *Precaution:* Never touch a germicidal lamp (not even for a second) while it is energized, because it will leave serious burns, even though the lamp may feel cool to the touch.

UNIT 21

Sunlamps

Since ultraviolet energy capable of causing reddening of the skin (sunburn) is provided by natural sunlight, devices designed to produce a similar energy are known as sunlamps.

Exposure of the skin to radiation from sunlamps results in the reddening of the skin a few hours after exposure and may last for several days. If the amount of energy is sufficient, sunburn is normally followed by tanning and may last for months.

Absorption of this type of energy by the skin also stimulates the formation of vitamin D—the only vitamin that can be produced by the human body. This type of energy is also valuable in the prevention of certain diseases as well as the common cold.

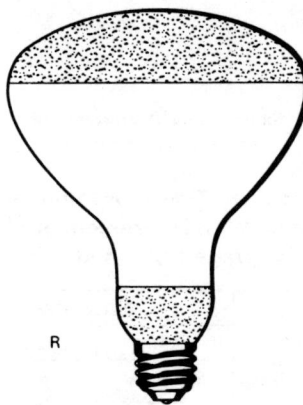

Fig. 21-1. A typical RS reflector sunlamp.

LAMP TYPES

All mercury lamps generate an ultraviolet energy which causes sunburn or suntan, and this type of lamp is the basis of two commonly used sunlamps, the RS (reflector sunlamp), and the fluorescent sunlamp.

RS REFLECTOR LAMP

The self-contained RS reflector sunlamp with built-in starter switch and filament ballast uses an R-40 bulb like the one used for standard reflector lamps (Fig. 21-1). It will operate satisfactorily on alternating-current circuits only on either 50-to 60-Hz current within the range of 110 to 125 volts. It takes approximately two minutes to reach full ultraviolet output on starting and approximately three minutes to cool before restarting.

The outer bulb of the RS lamp is made of Pyrex glass which withstands sudden changes in temperature without cracking and may be used in locations where it may be splashed with water. This glass absorbs and eliminates the unwanted ultraviolet rays but transmits those which cause suntan. This amount of energy is sufficient to produce a mild sunburn on untanned skin in 5 to 10 minutes; this is equivalent to 15 to 18 minutes of exposure to midsummer sun.

The life of the RS lamp on normal alternating-current circuits in household use has been estimated at approximately 1000 hours when operated at 5 or more hours per start.

FLUORESCENT SUNLAMPS

Fluorescent sunlamps are of the hot cathode preheat type and are available in 20- and 40-watt sizes (Fig. 21-2). They produce suntanning ultraviolet energy in the same way that fluorescent lamps produce light and, in appearance, are identical with standard fluorescent lamps of the same wattage.

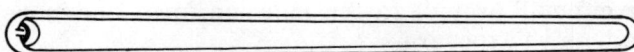

Fig. 21-2. A typical fluorescent sunlamp.

PRINCIPLES OF ILLUMINATION

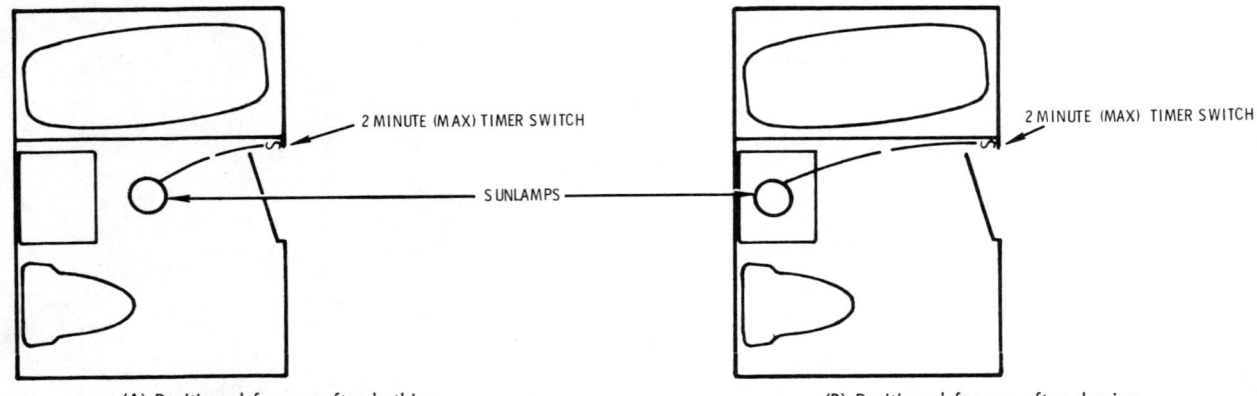

(A) Positioned for use after bathing.   (B) Positioned for use after shaving.

Fig. 21-3. Practical applications of the RS sunlamp in the bathroom.

The life of fluorescent sunlamps averages about five thousand 15-minute applications or 4000 hours at 6 hours per start. The effective ultraviolet output decreases throughout its life, with the most rapid drop occurring during its first 100 hours of operation.

PRACTICAL APPLICATION

The RS sunlamp can be conveniently screwed into an ordinary household socket without the necessity of any other equipment. It is often mounted in a fixed position, about two feet from the face, in residential bathrooms either over the shaving mirror (Fig. 21-3B) or in a position where one would normally dry off after bathing (Fig. 21-3A). Since the rays are directed downward, the eyebrows protect the eyes from radiation, and it is usually not necessary to wear goggles or glasses during exposure. However, the user should avoid looking directly into the lamp for any appreciable length of time.

Table 21-1. Exposure Time and Area Covered

Distance (ft)	Exposure Time (minutes)	Circular Area Covered (inches)
2	2	20
3	4	30
4	7	40

If the RS sunlamp is mounted 2 feet from the skin, it will cover a circular area of approximately 20 inches. The area coverage will increase about 10 inches for each additional foot of separation between lamp and skin, but the exposure time required will double (see Table 21-1).

Multiple mounting of two or more lamps in a group will provide faster, more uniform, and more extensive exposure.

Due to the shape of fluorescent lamps, a large part of the body can be uniformly tanned in a single exposure. Suitable fixtures are available for nearly any application. Fig. 21-4 illustrates one type of suitable mounting.

Exposure times for minimum perceptible erythema with fluorescent sun lamps are shown in Table 21-2.

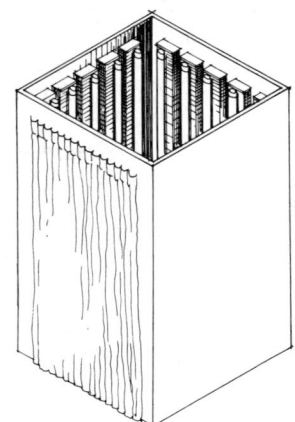

Fig. 21-4. A fixture using fluorescent sunlamps to tan entire body uniformly.

Table 21-2. Exposure Times for Minimum Perceptible Erythema With Fluorescent Sunlamps (Average Untanned Skin)

	Distance From Lamp		
	1 ft	2 ft	3 ft
Lamp	Exposure Time (minutes)	Exposure Time (minutes)	Exposure Time (minutes)
20-Watt Plus Reflector	3	7	14
20-Watt Bare	8	20	41
40-Watt Plus Reflector	2	4	6
40-Watt Bare	5	10	18

SUMMARY

- All mercury lamps generate ultraviolet rays which cause sunburn or suntan.

- A special kind of mercury lamp is the sunlamp which, by selective absorption of the glass bulb, limits the generated rays to those found in natural sunlight, causing sunburn but eliminating the unwanted ultraviolet rays.
- The self-contained RS reflector sunlamp is equipped with a built-in starting switch and filament ballast.
- The life of the RS sunlamp is rated at approximately one thousand 15-minute applications.
- Fluorescent sunlamps are manufactured in 20- and 40-watt sizes and are rated at approximately five thousand 15-minute applications.
- Fluorescent sunlamps must be provided with auxiliary equipment for starting and operation.

UNIT 22

Lamps for Use in Horticulture

For a number of years, many individual gardeners have used a combination of fluorescent and incandescent lamps to start their seedlings indoors in the early spring. Although seedlings become strong growing plants when exposed to these lights for several weeks, most such operations were on a very small scale. Now, with new and improved lamps and equipment, large commercial growers and farmers have started full-scale experimentation with most promising results.

ADVANTAGES OF ARTIFICIAL LIGHTING

The practice of growing plants in a controlled environment (growth rooms) where natural light is replaced by artificial light permits:

1. More predictable plant response.
2. More precise timing of crops.
3. Easier management.
4. The saving of labor.

Lighting in greenhouses before sunrise and after sunset extends the light period and decreases the normal growth time.

Lighting in greenhouses on dark or overcast days also extends the light period and decreases the normal growth time.

Lighting under benches can double growing areas within a greenhouse.

Table 22-1 will serve as a guide in determining a proper lighting layout for plant growth.

For example, it has been determined that a winter crop of pot chrysanthemums develops improved growing and flowering qualities when grown under GE Lucalox lamps the last three weeks of

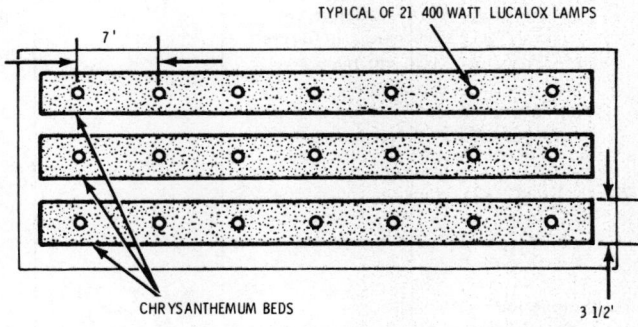

Fig. 22-1. A lighting system using 400-watt Lucalox lamps to grow chrysanthemums.

their growing cycle. During this three-week period, they are exposed to the lights 12 hours a day. See Fig. 22-1 for a suitable layout of such a lighting system using 400-watt Lucalox lamps. They are mounted 4 feet above the bed and produce about 500 footcandles—16 watts per square foot.

Table 22-1. The Requirements for Photosynthetic and Photoperiodic Lighting

Object of Lighting	Applications	Time Applied	Total Effective Light Period (hours)	Lamp (W/ft^2)	Light Sources	Luminaires
I. Photosynthetic **A. Supplementary** 1. Daylength extension	a. Seed germination, seedlings, bulb forcing b. Mature plants	4 to 10 hours before sunrise and/or after sunset	a. 12 to continuous b. 10 to continuous	a. 5 to 20 b. 10 to 40	Fluorescent, mercury & fluorescent-mercury lamps of various wattages with & without internal reflectors & used with or without 10 to 30% of installed watts of incandescent	Moisture-resistant luminaires of industrial or custom-made designs with mountings fixed or adjustable providing minimum interference with greenhouse routine & uniform light distribution
2. Dark day	As above	Total light period	As above 16	As above		
3. Night	As above	4 to 6 hours in middle of dark period		Same		
4. Under bench	As above	Total light period	10 to continuous	Same	Fluorescent lamps	Moisture-resistant, direct reflector units with mounting for uniform light distribution
B. Growth room 1. Professional horticulture	Seed germination, seedlings, cuttings, bulb forcing	Total light period	12 to continuous	5 to 30	Fluorescent lamps with or without 10 to 30% incandescent or combination of plant growth lamps	Industrial direct reflector luminaires which are moisture resistant & are mounted in a shelf arrangement
2. Amateur horticulture	Seed germination, seedlings, cuttings, bulb forcing, mature plants, etc.	Total light period	10 to continuous	5 to 30	Fluorescent lamps (plant growth lamps) with and without incandescent	As above
3. Experimental horticulture	All types of plant responses	Total light period	0 to continuous	0 to 140 & higher	Many types used to fit the requirements of tests. Generally fluorescent with 10 to 30% incandescent lamps or combination of plant growth and wide-spectrum lamps.	Custom built with minimum spacing for maximum light output of lamps with uniform light distribution
II. Photoperiodic **A. Supplementary** 1. Daylength extension	Long-day effect to prevent flowering of short-day plants and induce flowering of long-day plants	4 to 8 hours before sunrise and/or after sunset	14 to 16	.5 to 5	Fluorescent, fluorescent-mercury, and incandescent lamps	As for photosynthetic supplementary lighting
2. Night break	As above	2 to 5 hours in middle of dark period	14 to 16	.5 to 5	As above	As above
3. Cyclic	As above	1 to 4 seconds per minute, 1 to 4 or 10 to 30 minutes per hour	14 to 15	1 to 5	Mostly incandescent, or fluorescent lamps with flashing ballasts	As above

UNIT 23

Miscellaneous Applications

Earlier units of this book cover the basic procedures required to design lighting systems for a variety of seeking tasks; however, there are endless applications for the use of light which would be impossible to cover in this book. Light, for example, is used to dry grain stored in bins; automobile body shops use light to dry paint and lacquer; light is used to capture and trap insects; etc.

The following paragraphs will describe two unusual applications of light for the purpose of seeing.

LIGHTING WITH SUNLIGHT

A huge multistory office building, located in the business section of a familiar city, stands lifeless in the cool morning hours before sunrise. The outline of a motionless apparatus, resembling a large mirror, can be seen on the roof as it stands out against the brighter moonlit sky. Suddenly the warmth of the first rays of morning sun strike an ultrasensitive thermostat located near the apparatus. This energizes an electric motor which swings the westward-facing mirror quickly over to the east. From that point the mirror is pivoted by a second motor, controlled by three other sun-operated thermostats, until the solar image is reflected in the center of the steel mirror. Then, almost instantaneously, every room in the huge building is showered with bright diffused sunlight.

Above the mobile mirror is a large fixed mirror facing downward at an angle over a square shaft in the center of the building. The rays from the mobile mirror reflect on to the fixed mirror, which deflects them in a powerful 32,000-candlepower beam straight down the shaft to the basement. Narrow shafts of light are collected from the main beam and reflected by mirrors from room to room through small apertures at ceiling level on each floor. Regardless of how far the rays must travel, there is no loss of light intensity.

If the sun is obscured by a passing cloud, the thermostat immediately stops the motor, at the same time switching on the auxiliary electric lighting in the rooms below. When the sun comes out again, the starter-motor again goes into action, swinging the mirror forward until it catches up with the new position of the sun, which, in turn, switches off the electric lights in the rooms below.

General illumination within the rooms is provided by the rays of the sun passing through diffused glass panels in the ceiling, while direct beams are focused at will on desks and for similar seeing tasks.

This sun-tapping apparatus comes complete with several very ingenious accessories. An automatic mirror-wiper prevents the upturned mirror from getting dirty. A motor-operated arm automatically sponges and polishes the mirror once a day as it swings from west to east to catch the morning sun. The pilot-thermostat controls the mirror motor, causing it to move at the same speed as the sun.

Is this a newly introduced apparatus? A proposed device for the future? Neither! It is known as the Arthel Heliostat and was successfully used for the interior lighting of buildings in the 1930s. However, the full-time use of electrical energy offered a less expensive method of lighting buildings initially, and the Heliostat lost ground and was eventually discarded.

Depending on the latitude in which it was used, it offered a saving in electrical energy of between 35 and 80% and with America now facing the present energy shortage, it may be wise to take another look at this apparatus which taps our first light source—the sun.

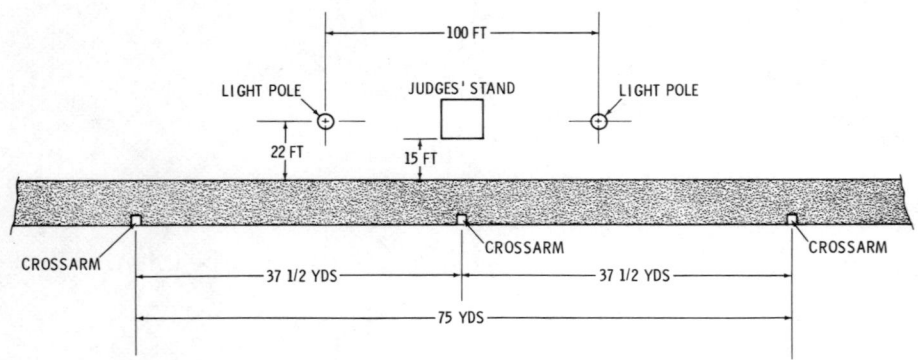

Fig. 23-1. A site plan of the major portion of the jousting area.

LIGHTING FOR JOUSTING

Older than the Kentucky Derby, older than baseball, born in the Middle Ages and nursed by chivalry, that's the story of jousting, a test of horsemanship, balance and marksmanship. To the command, "Charge, Sir Knight!" renewal of America's oldest consecutively run sporting event takes place each year on the third Saturday in August at Natural Chimneys, Mt. Solon, Virginia, in a revival of medieval pageantry.

Originally a game in which death frequently claimed the loser, present-day riders test their skill at plucking three steel rings from crossbars with their lances, instead of sending the same through the breast plate of an iron-clad foeman. Jousting, however, has lost none of its flavor in its contemporary dress. Modern "knights" in bright garb charge down a 75-yard course in not more than eight seconds, sending their precisely maneuvered lances through the ultrasmall steel rings.

The sport is also popular in other locations in the eastern United States, but no meet is as large, as old, or as rich in tradition as the Mt. Solon event. This is why the Upper Valley Regional Park Authorities of Virginia decided to improve the old jousting facilities when they purchased the grounds for a state park recreation facility.

Director Jack Vaughan then contacted Dwight E. Miller, AIA Architect to design the various recreation facilities for the park. Our firm was then asked to design the mechanical and electrical systems for the entire park and of the various facilities contained in this 134-acre park; the small jousting area offered the greatest challenge.

Prior to this project, I had no idea that jousting tournaments were still held anywhere in the world, much less at an area not 50 miles from my home. However, I soon found out that I was not alone, as several inquiries to sports-lighting manufacturers and lighting associations turned up no information on lighting design for this sport. Some of the people I talked with were probably wondering just what institution I had escaped from. Consequently, it was evident that I would have to start from scratch.

First, an analysis of the old existing lighting was made. This lighting system consisted of two 175-watt mercury lighting fixtures mounted on one centrally located 30-foot pole, and each ring-holding crossarm had a shallow dome reflector with a 100-watt inside frosted lamp mounted on a ½-inch conduit approximately 12 inches out from the crossarm. While this lighting arrangement seemed to have been adequate, many of the jousters

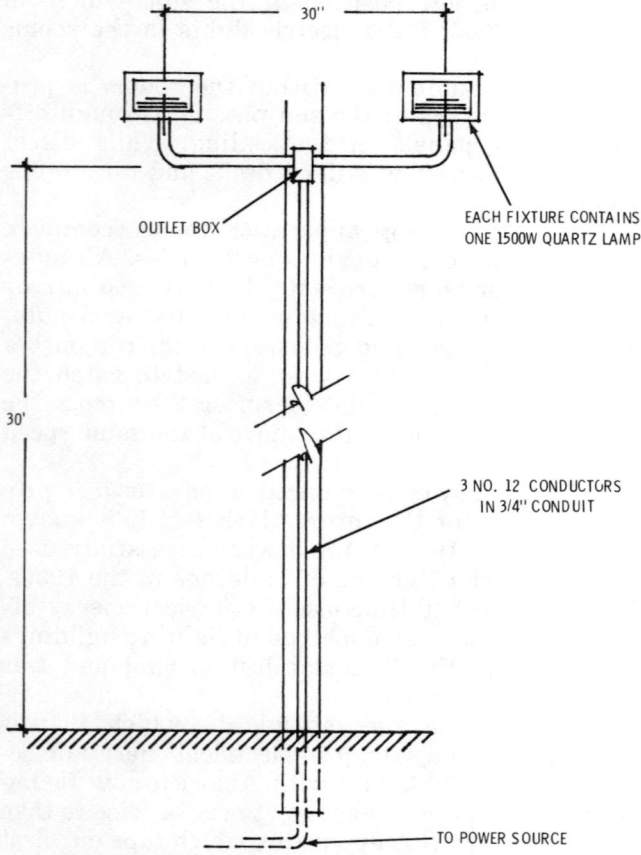

Fig. 23-2. General lighting arrangement of the jousting area.

MISCELLANEOUS APPLICATIONS

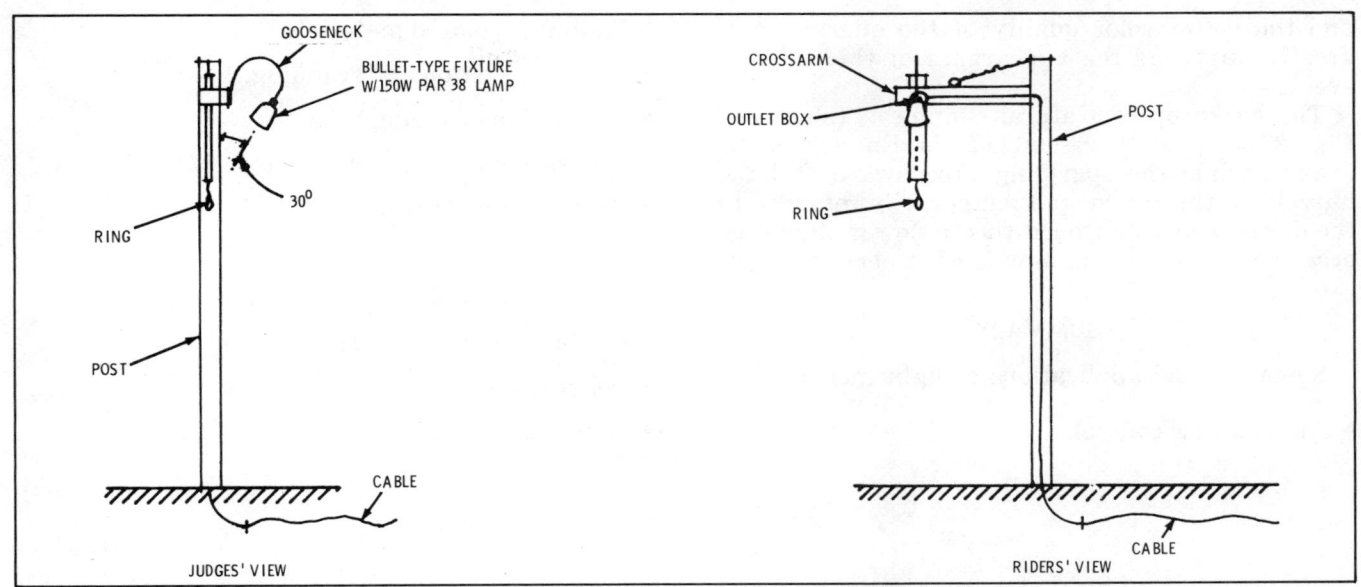

Fig. 23-3. Lighting the ring by using 150-watt PAR-38 lamps.

claimed that it left much to be desired; they especially disliked the poor color of the standard mercury lamps. With just this data, I began the lighting design.

Fig. 23-1 shows the site plan of the major portion of the jousting area. The course, or "route of riders," is 75 yards long, and the three crossarms containing the rings are 37½ yards apart. Two 30-foot wood poles were set 22 feet back from the course and 50 feet from the center and to each side of the judges' stand. The type of lamps, fixtures, and mounting details are illustrated in Fig. 23-2. This general lighting arrangement gave an average illumination level of not less than 5 footcandles anywhere on the course, and provided sufficient general illumination for the riders, viewers, and judges.

Each of the three crossarms had a bullet-type fixture mounted on a gooseneck from the crossarm, as illustrated in Fig. 23-3. Each contained a 150-watt PAR-38 lamp. This arrangement gave approximately 100 footcandles of illumination directly on the ring without objectionable glare to either horse or rider.

In August 1972, the first meet using this new lighting system was held, and nearly all of the contestants seemed to be pleased with the new system. Greater speed and better points were made,

Fig. 23-4. The backdrop of Natural Chimneys.

and the better color quality of the quartz lamps greatly improved the appearance of this colorful event.

The backdrop of Natural Chimneys (shown in Fig. 23-4) greatly resembled old English castle towers, while the sparkling greensward and the huzzahs of the crowd—all temporarily returned to the days when knighthood was in flower, made the occasion unique in the annals of American sport.

SUMMARY

Some unusual applications of light include:

- Photoelectric control.
- Fading of colored materials.
- Insect attraction and trapping.
- Production of vitamin D.
- Prevention and cure of rickets.
- Photochemical actions.
- Microorganism growth control.
- Radiant heating and heat therapy.
- Production drying, softening, and heating.
- Dehydration.
- Comfort heating.

SECTION VII

LIGHTING COST ANALYSIS

UNIT 24

The Cost of Lighting

Total cost of lighting can be computed by assembling all the factors of both fixed and operating charges which apply to any given installation. The total thus obtained may serve either as a comparison of the cost of light with other elements of production or as a comparison of various lighting systems.

A typical initial cost of a 100-footcandle lighting installation, including installation and branch circuit wiring cost, is approximately $1.00 per square foot for industrial applications to over $1.50 per square foot for commercial applications. These figures are approximately correct whether the system is fluorescent, mercury, or incandescent. In small rooms, this cost may go as high as $3.00 per square foot for 100 footcandles, especially where several decorative-type fixtures are used.

Table 24-1. Cost-of-Light Comparison

	200-Watt Incandescent	40-Watt Fluorescent
Watts	200	40
List Price	$0.50	$1.50
Rated Average Life	750 hours	25,000 + hours
Estimated Mean Lumens	3710	1765
Replacement Labor	$2.00/lamp	$2.00/lamp
Electricity Rate	$0.02/kwh	$0.02/kwh

For example, a 100-footcandle lighting installation in a 10,000-square-foot factory would cost approximately ($1.00 × 10,000 square feet) $10,000.

OWNING COST

The owning cost of most commercial lighting systems is approximately 30¢ per square foot per

Table 24-2. GE Q250PAR38 Operating Cost-of-Light Comparisons

RECOMMENDED SUBSTITUTION OF Q250PAR38/FL to achieve equal maintained light	ONE Q250PAR38/FL for	TWO 150R/FL or	TWO 150PAR/FL
LAMPS			
A. Lamp Cost (List)	$11.25	$1.49	$2.45
B. Lamp Replacement Per Socket Per Year (Burning Time per yr—4000 hour/lamp life)	1.0	2.0	2.0
C. Number of Sockets	1.0	2.0	2.0
D. ANNUAL LAMP COST (A×B×C)	$11.25	$5.96	$9.80
LABOR			
E. Labor Charge Per Lamp Replacement (Est. Avg.)	$3.00	$3.00	$3.00
F. ANNUAL LABOR COST (B×C×E)	$3.00	$12.00	$12.00
ELECTRICITY			
G. Kilowatt Hours Per Socket Per Year (2c/KWHR × Burning Time/1000)	1000 KWH	600 KWH	600 KWH
H. ANNUAL ELECTRICITY COST (Energy Cost × C × G)	$20.00	$24.00	$24.00
TOTAL			
I. ANNUAL COST-OF-LIGHT (D+F+H)	$34.25	$41.96	$46.80
ANNUAL SAVINGS POSSIBLE WITH Q250PAR38/FL		$7.71	$11.55

Table 24-2. GEQ250PAR38 Operating Cost-of-Light Comparisons—cont

RECOMMENDED SUBSTITUTION OF Q250PAR38/SP to achieve equal spotlighting (approx.)	ONE Q250PAR38/SP for	THREE 150PAR/SP or	TWO 300R/SP
LAMPS			
A. Lamp Cost (List)	$11.25	$2.45	$2.30
B. Lamp Replacements Per Socket Per Year (Burning Time per yr-4000 hour/lamp life)	1.0	2.0	2.0
C. Number of Sockets	1.0	3.0	2.0
D. ANNUAL LAMP COST (A×B×C)	$11.25	$14.70	$9.20
LABOR			
E. Labor Charge Per Lamp Replacement (Est. Avg.)	$3.00	$3.00	$3.00
F. ANNUAL LABOR COST (B×C×E)	$3.00	$18.00	$12.00
ELECTRICITY			
G. Kilowatt Hours Per Socket Per Year (Lamp Wattage × Burning Time/1000)	1000 KWH	600 KWH	1200 KWH
H. ANNUAL ELECTRICITY COST (2c/KWHR ×C × G)	$20.00	$36.00	$48.00
TOTAL			
I. ANNUAL COST-OF-LIGHT (D+F+H)	$34.25	$83.40	$69.20
ANNUAL SAVINGS POSSIBLE WITH Q250PAR38/FL		$49.15	$34.95

Courtesy General Electric Co.

year per 100 footcandles. This is assuming the initial cost is amortized over a period of 10 years at 8% interest.

OPERATING COST

Typical operating cost of commercial lighting, which includes electricity, cleaning and relamping, will cost between 30 and 35¢ per square foot per year per 100 footcandles.

TOTAL COST

The total annual owning and operating cost for commercial lighting systems will normally run between 60 and 65¢ per square foot per year per 100 footcandles.

A typical cost analysis form for comparing cost of various lighting systems is shown in Fig. 24-1.

An actual comparative economic study, as prepared by Crouse-Hinds/Revere lighting manufacturers, is given in Fig. 24-2.

Table 24-3 lists a cost-of-light comparison as prepared by General Electric—typical 200-watt incandescent lighting system versus deluxe warm white 40-watt fluorescent system.

Operating cost-of-light comparisons of the General Electric Q250 PAR-38 can be found in Table 24-4.

COST OF LIGHT COMPARISON

A cost of light comparison was made by the General Electric Company (Table 24-1) comparing the 200-watt incandescent lighting system to the deluxe warm white, 40-watt fluorescent lighting system. The results of this comparison are as follows:

The number of fluorescent lamps needed to match the incandescent light level are:

$$\text{Mean Lumens} = \frac{3710}{1765}$$
$$= 2.1 \text{ lamps}$$

The number of incandescent lamp replacements needed to match the fluorescent lamp life is:

$$\text{Average life hours} = \frac{25{,}000}{750}$$
$$= 33.3 \text{ lamps}$$

To find the cost ratio of the incandescent lighting system to the fluorescent lighting system, find the total cost of each system and then divide the cost of the fluorescent system into the cost of the incandescent system. First, find the cost of the incandescent system.

Replacement lamps:
  33.3 lamps × $0.50   = $ 16.65
Replacement labor:
  33.3 lamps × $2.00   = $ 66.60
Electricity: 1 lamp × 200 watts ×
  25,000 hours × $0.02 per kwh   = $100.00
                              Total $183.25

Second, find the cost of the fluorescent system.

Replacement lamps:
  2.1 lamps × $1.50   = $ 3.15
Replacement labor:
  2.1 lamps × $2.00   = $ 4.20
Electricity: 2.1 lamps × 40 watts ×
  25,000 hours × $0.02 per kwh   = $42.00
                              Total $49.35

$$\text{Cost ratio} = \frac{\$183.25}{\$49.39}$$
$$= 3.7:1$$

TYPICAL COST-ANALYSIS FORM

	Lighting System Number	1	2
Description of Lighting Systems	1. Type of lamp (filament, mercury, preheat fluorescent, slimline, etc.)	----	----
	2. Lamp description	----	----
	3. Type of luminaire	----	----
	4. Number of lamps per luminaire	----	----
Basic Data	5. Rated initial lumens per luminaire	----	----
	6. Lamp life	----	----
	7. Watts per luminaire (including auxiliary)	----	----
	8. Coefficient of utilization	----	----
	9. Maintenance factor	----	----
	10. Number of luminaires	----	----
	11. Average footcandles maintained	----	----
	12. Energy rate ($ per kwh)	----	----
	13. Estimated burning hours per year	----	----
Initial Cost	14. Net luminaire cost (each)	----	----
	15. Net additional accessory cost per luminaire	----	----
	16. Estimated wiring and installation cost per luminaire. This includes all wiring materials and fixtures and the labor required to install them. Panels, feeders, and transformers, if necessary, should be proportioned in this cost.	----	----
	17. Net initial lamp cost each (list less ___% + tax)	----	----
	18. Net initial lamp cost per luminaire (4 x 17)	----	----
	19. Total initial cost per luminaire (14 + 15 + 16 + 18)	----	----
	20. Total initial cost (10 x 19)	----	----
Annual Fixed Charges	21. Initial cost per luminaire less lamps (14 + 15 + 16)	----	----
	22. Total initial cost less lamps (10 x 21)	----	----
	23. Annual fixed charges (___% of 22) The proportion of the initial expenditure to be written off each year is dependent upon the nature of the business. In considering industrial plants where luminaire styles change infrequently, 10% or less may be used. The motifs of commercial establishments change frequently, and the depreciation factor may approach 20%. To this should be added the interest, taxes, insurance, etc. These will usually vary between 5 and 10% of total fixed charges. Purchaser is consulted.	----	----
Annual Operating Costs	24. Annual number of lamp replacements (4 x 10 x 13 ÷ 6)	----	----
	25. Annual cost of replacement lamps (17 x 24)	----	----
	26. Annual cost of replacement parts (starters, etc.)	----	----
	27. Total annual maintenance material cost (25 + 26)	----	----
	28. Estimated labor cost to replace one lamp	----	----
	29. Total labor cost to replace lamps (24 x 28)	----	----
	30. Estimated cleaning cost per luminaire	----	----
	31. Number of cleanings per year	----	----
	32. Annual cleaning cost (10 x 30 x 31)	----	----
	33. Total annual maintenance labor cost (29 through 32)	----	----
	34. Total annual maintenance cost (27 + 33)	----	----
	35. Annual energy cost (7 x 10 x 12 x 13 ÷ 1000)	----	----
	36. Total annual operating cost (34 + 35)	----	----
Total and Relative Costs	37. Total annual cost (23 + 36)	----	----
	38. Relative annual cost	----	----
	39. Annual cost per footcandle (37 minus 11)	----	----
	40. Relative annual cost per footcandle	----	----

Fig. 24-1. A typical cost analysis form.

Comparative Economic Study

		System A Mercury	System B Metallic Additive
Initial Cost			
1.	Material:		
	a. Floodlights	$4032.00	$3456.00
	b. Ballasts	(incl w/fldts)	(incl w/fldts)
	c. Lamps	240.96	334.80
	d. Poles	1176.00	948.00
	e. Cable, switchgear, etc.	300.00	225.00
	f. Total material cost	$5748.96	$4963.80
2.	Labor:		
	a. Labor rate/hour	$7.00	$7.00
	b. Average installation time (hours)	160.00	120.00
	c. Total installation labor costs (2a x 2b)	$1120.00	$840.00
3.	Total Initial Cost (1f + 2c):	$6868.96	$5803.80
4.	Annual owning cost (15% of item 3 less lamp cost—depreciation 10%; interest, taxes and insurance 5%)	$994.20	$820.35
Annual Operating Costs			
5.	Annual Energy Costs:		
	a. Burning hours/yr	2000	2000
	b. Watts consumed/lamp	x 440	x 440
	c. Number of lamps	x 32	x 24
	d. Utility rate (divide $ rate per kwh by 1000 for $ rate per wh)	$ .000025	$ .000025
	e. Total annual energy cost (5a x 5b x 5c x 5d)	$704.00	$528.00
6.	Annual Lamp Costs:		
	a. Burning hours/yr	2000	2000
	b. Rated lamp life (hours)	16000	7500
	c. Relamps/yr (6a/6b)	.125	.267
	d. Lamp cost	x $7.35	x $13.95
	e. No. lamps on job	x 32	x 24
	f. Total lamp cost/yr (6c x 6d x 6e)	$29.4	$89.39
7.	Annual Relamp Labor Costs:		
	a. Relamp/yr	.125	.267
	b. Time to relamp (hours)	8	6
	c. Relamp labor rate	$7.00	$7.00
	d. Total annual relamp labor costs (7a x 7b x 7c)	$7.00	$11.21
8.	Cost of cleaning (frequently-annual, 1hour/pole, $7.00/hour)	$56.00	$42.00
9.	Total Annual Operating Cost (5e + 6f + 7d + 8)	$796.14	$670.60
10.	Total Annual Cost (4 + 9)	$1790.34	$1490.95

CONCLUSION: Since the total annual cost of the metallic additive system is less than the mercury, the metallic additive GAL-4-M system should be used.

Fig. 24-2. A comparative economic study.

Courtesy Crouse-Hinds Co.

SECTION VIII

LIGHTING CONTROLS AND WIRING

UNIT 25

The Control of Lighting

Many lighting control devices have been developed since Edison's first lamp. They have been designed to make the best use of the lighting equipment provided by the lighting industry. These include: the quite complex control devices for signs and advertising as described in Unit 19; automatic timing devices for street and outdoor lighting; dimmers for theatrical, studio, and church lighting in order to achieve certain objectives; and of course the common light switch used in nearly every home in the nation. In all fields of lighting, the usefulness and the convenience of lighting installations can be greatly improved by the proper application of control devices.

THE SWITCH

A switch, for our purposes, is a device used on branch circuits to control lighting. Switches fall into the following basic categories:

1. Snap-action switches.
2. Mercury switches.
3. Quiet switches.

Snap-Action Switches

A single-pole snap-action switch consists of a device containing two stationary current-carrying elements, a moving current-carrying element, a toggle handle, a spring, and a housing. When the circuit is open, as in Fig. 25-1A, no current can pass, and the light is off. When the moving element is closed, by manually flipping the toggle handle, the circuit is closed and the light will burn (Fig. 25-1B).

Mercury Switches

Mercury switches consist of a sealed capsule containing mercury, as illustrated in Fig. 25-2. Inside the capsule are contacting surfaces "A" and "B," which may be part of the wall of the capsule. The switch is operated by means of a handle which moves the capsule.

As shown in Fig. 25-2, the capsule is tilted so that the mercury "C" has collected at one end of the capsule. Here, it bridges two contact points, "A" and "B," to complete the circuit and light the lamp. However, if the capsule is tilted the opposite

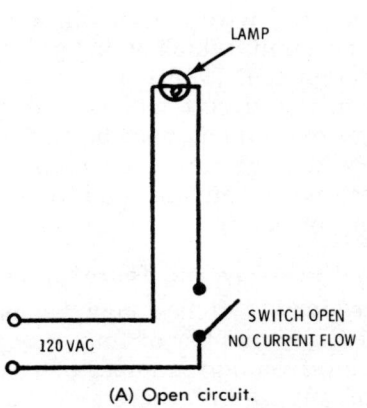
(A) Open circuit.

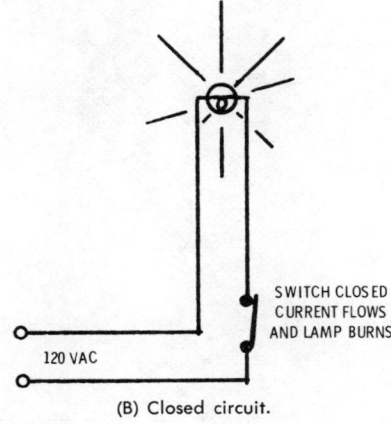
(B) Closed circuit.

Fig. 25-1. A single-pole snap-action switch circuit.

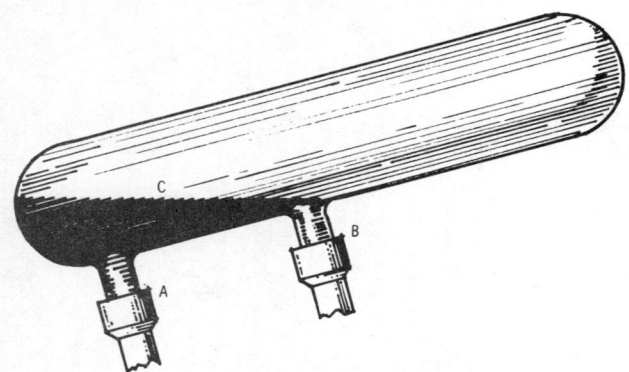

Fig. 25-2. The operation of a mercury switch illustrated.

way, the circuit between contacts "A" and "B" will not be completed, and the lamp will be off. Mercury switches offer the ultimate in silent operation and are recommended for use in high quality homes and where the "clicking" of a light switch may be annoying.

Quiet Switches

The quiet switch is a compromise between the snap-action switch and the mercury switch. Its operation is much quieter than the snap-action switch, yet it is not as expensive as the mercury switch.

The quiet switch consists of a stationary contact and a moving contact that are close together when the switch is open. Only a short, gentle movement is applied to open and close the switch, producing very little noise. This type of switch may be used only on alternating current, since the arc will not extinguish on direct current.

The quiet switch is the most commonly used switch for modern lighting practice. These switches are common for loads from 10 to 20 amps, in single-pole, three-way, four-way, etc., configurations. Fig. 25-3 illustrates a modern quiet switch.

Many other types of switches are available for lighting control. One type of switch used mainly in the home is the door-actuated type which is generally installed in the door jamb of a closet to control a light inside the closet. When the door is open, the light comes on; when the door is closed, the light goes out. The light in your refrigerator is probably controlled by a door switch.

The Despard switch is another special switch. Because of its small size, up to three may be mounted in a standard single-gang switch box. Waterproof switches are made for outdoor use. Combination switch-indicator light assemblies are also available for use where the light cannot be seen from the switch location, such as in an attic or garage. Switches are also made with small neon lamps in the handle which light when the switch is off. These low-current-consuming lamps make the switches easy to find in the dark.

Three-Way Switches

Three-way switches are used to control one or more lamps from two different locations, such as at the top and bottom of stairways. The connection of three-way switches is illustrated in Fig. 25-4. By means of the two three-way switches, it is possible to control the lamp from two locations. By tracing the circuit, it may be seen how these three-way switches operate. Two wires are connected to the 120-volt line; one wire is connected directly to the lamp, while the other continues on to one of the three-way switches. Now if both handles of the three-way switches are in the "up" position as in Fig. 25-4, the current will pass through the top "traveler" wire and on through the other switch to the lamp, which will light because the circuit is completed. If either of the handles are turned down, the circuit will be broken, and the lamp will go out. But it may be turned on again at either switch, as any change of position of either switch will complete the circuit thus turning the lamp on again.

Combination Three-Way and Four-Way Switches

Two three-way switches may be used in conjunction with any number of four-way switches to control a lamp from any number of positions. The actuation of any one of these switches will turn the light on or off.

Courtesy Pass & Seymour, Inc.
Fig. 25-3. A quiet switch.

THE CONTROL OF LIGHTING

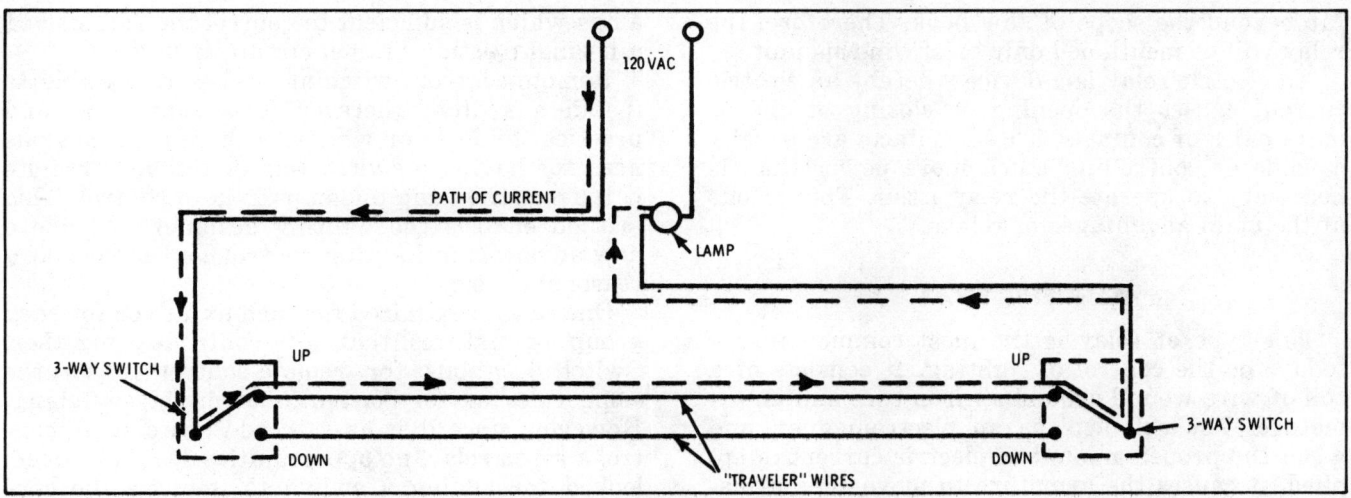

Fig. 25-4. A typical three-way switch connection.

Photoelectric Switch

The chief applications of the photoswitch are to control street lighting, sign lighting, and certain outdoor floodlighting installations, especially the "dusk-to-dawn" lights found in suburban areas.

This interesting switch has an endless number of possible uses and is a great tool for the modern lighting designer. Fig. 25-5 illustrates one type of photoswitch.

RELAYS

Next to switches, relays play the most important part in the control of light. However, the design and application of relays is a study in itself, and

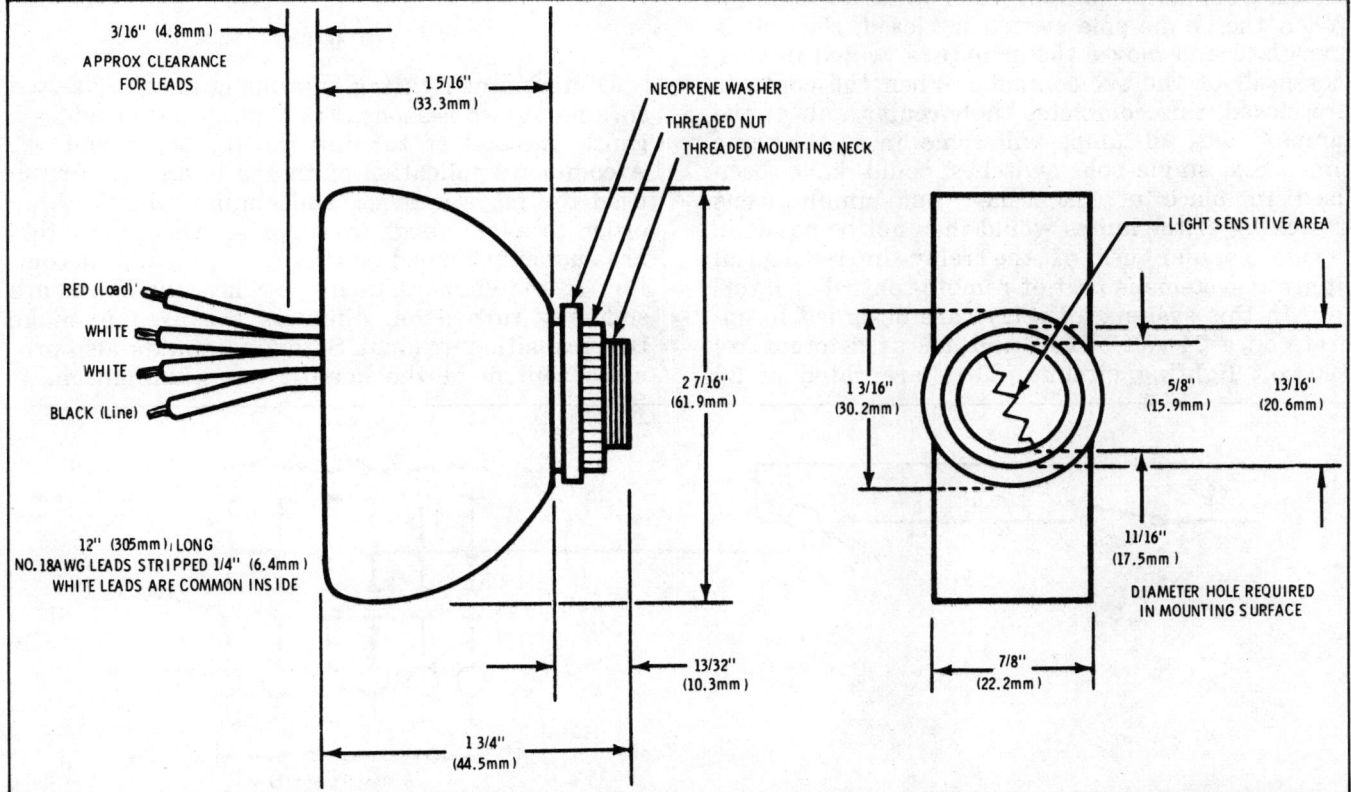

Courtesy Tork Time Controls

Fig. 25-5. Data for a photoelectric switch.

149

PRINCIPLES OF ILLUMINATION

far beyond the scope of this book. Therefore, the relay will be mentioned only briefly in this unit.

An electric relay is a device whereby an electric current causes the opening or closing of one or more pairs of contacts. These contacts are usually capable of controlling much more power than is necessary to operate the relay itself. This is one of the main advantages of relays.

MAGNETIC RELAY

This type of relay is the most common in use today for the control of lighting. It consists of a coil of wire wound around an iron core and an armature. The coil acts as an electromagnet, and when the proper amount of electric current is applied, it causes the armature to move. The armature, in turn, moves contact points together or apart, depending on how the relay is constructed. When the power is disconnected from the coil, a spring returns the armature to its original position. The following example demonstrates how a relay and a 15-amp single-pole switch can be used to control 60 amps of lighting.

Fig. 25-6 shows a one-line diagram of a relay controlled by a single-pole switch. The relay operates six contacts, each of which is connected to a circuit with lamps totaling 1200 watts (10 amps). When the single-pole switch is closed, the coil is energized and moves the armature, which in turn closes all of the six contacts. When the contacts are closed, this completes the circuit to all of the lamps. Thus, all lamps will come on at the same time. Six single-pole switches could have been used in place of the relay, but simultaneous switching of the lamps would then not be possible.

One popular use of the relay in residential lighting systems is that of remote-control of lighting. In this system, all relays are designed to operate on a 24-volt circuit and are used to control 120-volt lighting circuits. They are rated at 20 amps which is sufficient to control the full load of a normal lighting branch circuit, if desired.

Remote-control switching makes it possible to install a switch wherever it is convenient and practical to do so or wherever there is an obvious need for having a switch—no matter how remote it is from the lamp or lamps it is to control. This method enables the lighting designer to achieve new advances in lighting control convenience at a reasonable cost.

One relay is required for each fixture or for each group of fixtures that are controlled together. Switch locations for remote-control follow the same rules as for conventional direct switching. However, since it is easy to add switches to control a given relay, no opportunities should be overlooked for adding a switch to improve the convenience of control.

Remote-controlled lighting also has the advantage of using selector switches at certain locations. For example, selector switches located in the master bedroom or in the kitchen of a home enable the owner to control every lighting fixture on the property from this location. The selector switch may turn on and off an outside or other light which customarily would be left on until bedtime and which might otherwise be forgotten.

DIMMERS

Dimming of lighting systems is usually desired for one of two reasons. First, dimming provides a gentle method of turning the lights on and off. A common application of this is in an auditorium used for plays, movies, and similar functions. In order to avoid shock or surprise when the lights are suddenly turned on or off, or to avoid discomfort to the dark-adapted eye when the lights are suddenly turned on, dimmers are used to make this transition gradual. Second, dimming also provides control of the quantity of illumination. It

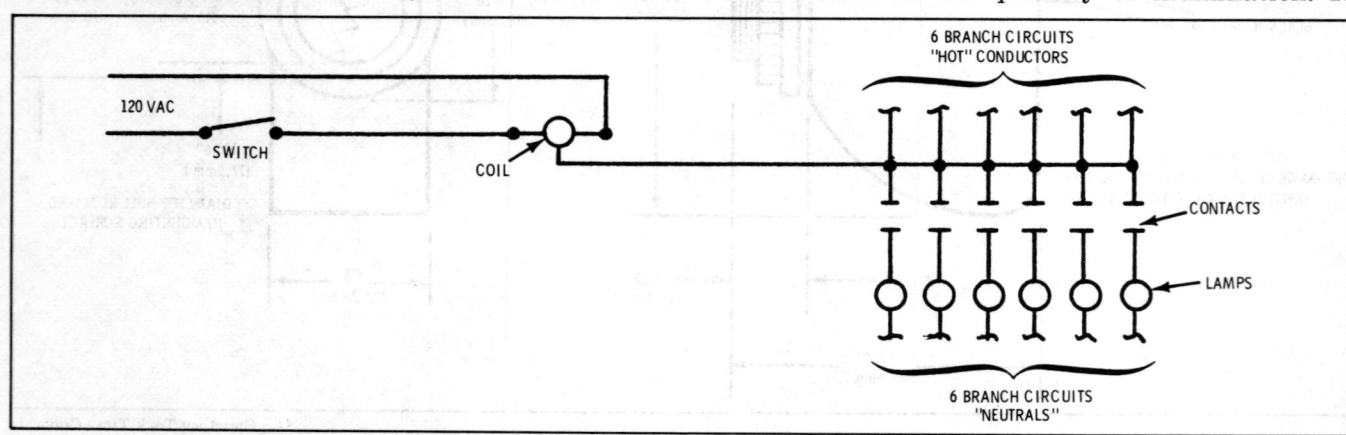

Fig. 25-6. Lighting controlled by a magnetic relay.

may be done to create various atmospheres and moods or to blend certain lights with others for various lighting effects.

Rheostat Dimmer

The oldest type of dimming equipment is the rheostat. This device was placed in series with an incandescent lamp circuit to vary the voltage which would reach the lamps. A dial on the rheostat varied the resistance within the circuit, which in turn changed the voltage (usually 0 to 120 volts). The less voltage reaching the lamps, the less light the lamps would give off.

The rheostat dimmer had its disadvantages and brought about the use of variable autotransformers for the most popular method of dimming ac incandescent lamps. The variable autotransformer controls voltage from full line voltage to zero, or even above line voltage by means of an extension winding.

Courtesy The Superior Electric Co.
Fig. 25-7. A residential-type dimmer switch.

For large installations, several autotransformers are used and are usually installed in banks and have control handles for the manual operation of individual dimmers or of a group of dimmers. Such banks are suitable for small stages, church auditorium lighting, as well as for general lighting control. However, for very large installations with several circuits, the banks of variable autotransformers are motor driven. Fig. 25-7 illustrates a residential-type dimmer switch for dimming up to 600 watts of lamps. Also Fig. 25-8 shows a bank of autotransformers equipped with remote positioners.

Dimming of Fluorescent Lamps

The advent of the rapid-start principle of fluorescent lamp operation made possible the dimming of fluorescent lamps. A cathode-heating supply current is provided with a constant voltage, while at the same time the arc passing through the tube is varied to produce dimming. In order to maintain this constant cathode-heating current while varying the lamp current, a special ballast transformer is necessary. It should be connected to the dimmer circuit as shown in Fig. 25-9.

The diagram in Fig. 25-9 shows that 120-volt line voltage is fed into the primary of the ballast transformer which produces constant voltage on the secondary winding connected to the cathode of the fluorescent lamp. The current which passes the length of the lamp, however, must pass through the main secondary of the ballast transformer and also through the dimmer. Leads connect to other ballasts (controlling other lamps) in the system. Dimming ballasts of this type are commonly available for 40-watt rapid-start fluorescent lamps.

FLASHING OF LAMPS

A flashing light is one of the best methods of getting attention, since it combines both brightness and motion. Its apparent motion or action has more attraction value for the human eye than any of the steadily glowing light sources. For these reasons, the application of flashing lights is especially desirable for advertising purposes. Economy is another reason for flashing. The time during which there is no light in a flashing action results in savings in electric power consumption and also increases the lamp life.

Incandescent, fluorescent, and neon lamps are the most-used types for flashing lights. The flashing may be accomplished by bimetallic thermal flashers but are most often flashed by means of motor-driven cam-actuated contact flashers. According to the effects produced and according to the actions performed, the basic types of flashers are:

1. Off-on.
2. Alternate speller.
3. Twinkler.
4. Border-chaser.

Used by themselves, these flashers can turn lights off and on, spell out words, cause a lamp to twinkle, or cause the border-chasing effect around a sign. By combining two or more types of flashers, the sign designer can obtain an almost endless number of other attention-getting effects.

PRINCIPLES OF ILLUMINATION

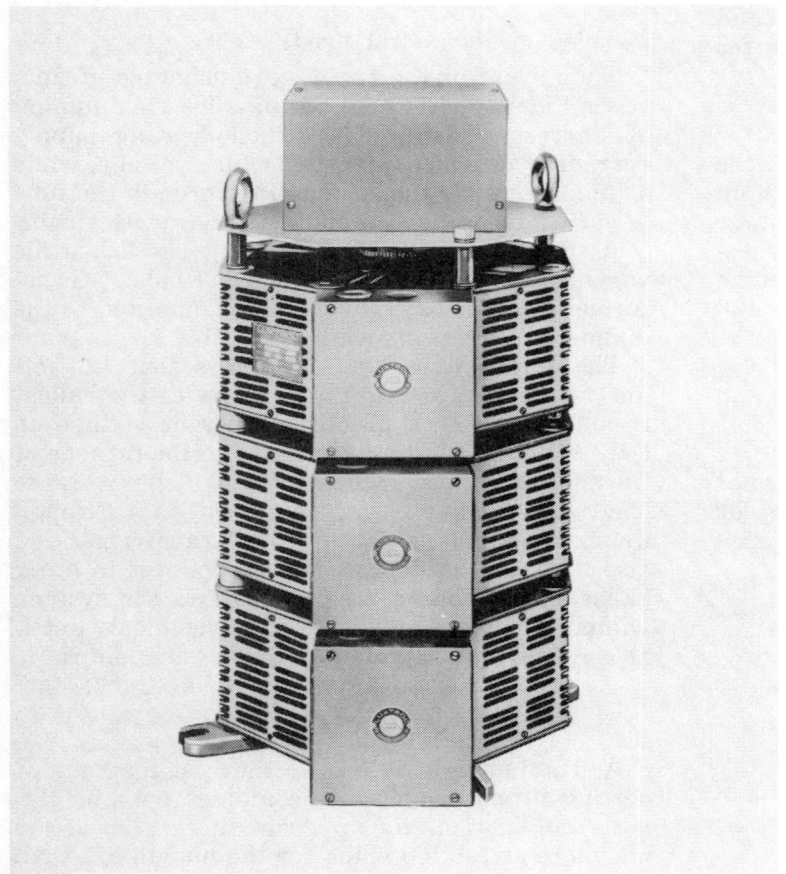

(A) Autotransformers.

(B) Remote positioners.

Courtesy The Superior Electric Co.

Fig. 25-8. A bank of autotransformers with remote positioners.

THE CONTROL OF LIGHTING

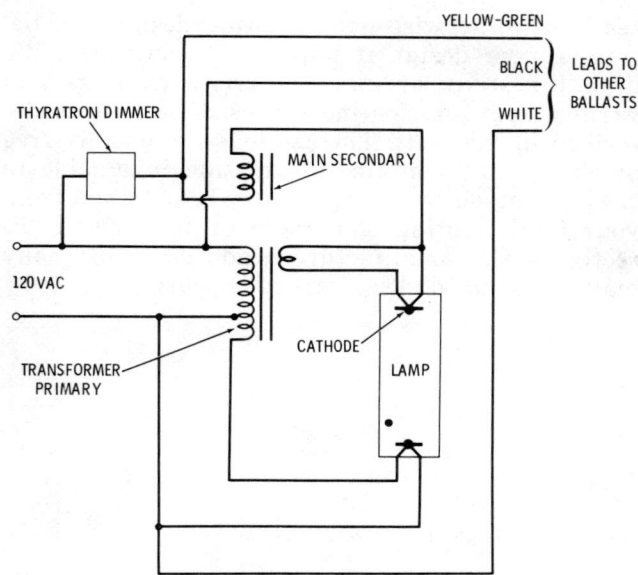

Fig. 25-9. A dimmer circuit using a special ballast transformer.

Off-On Flasher

The off-on flasher, as the name implies, turns the lights off and on at various intervals, producing both brightness and motion and conserving electrical energy—the amount depending on the time interval between flashing.

Speller Flasher

Speller flashers may be used to spell out the name of a product, manufacturer, agency, etc. By lighting sections of the lamps in a particular order, the effect of script writing may be accomplished. After the word is spelled out, it stays steadily lit for a few seconds, goes off, then starts the spelling action again. This is probably the most common type of flasher. This type of flasher can also be modified to perform many different actions, such as off-on, spelling, color-changing, and animation of figures.

Different effects are obtained by changing the opening and closing time of the circuit contacts.

Flashing Fluorescent Lamps

Rapid-start fluorescent lamps can be flashed without damage to the electrodes, provided these electrodes are kept heated to the right temperature at all times by using a special flasher ballast.

A typical two-lamp flasher-ballast circuit is shown in Fig. 25-10.

The flasher circuit is connected to the 120-volt line and continues to other ballasts (which control other lamps). The two fluorescent lamps are connected to the flasher ballasts, and the flashing action is controlled from the flasher as indicated. No-

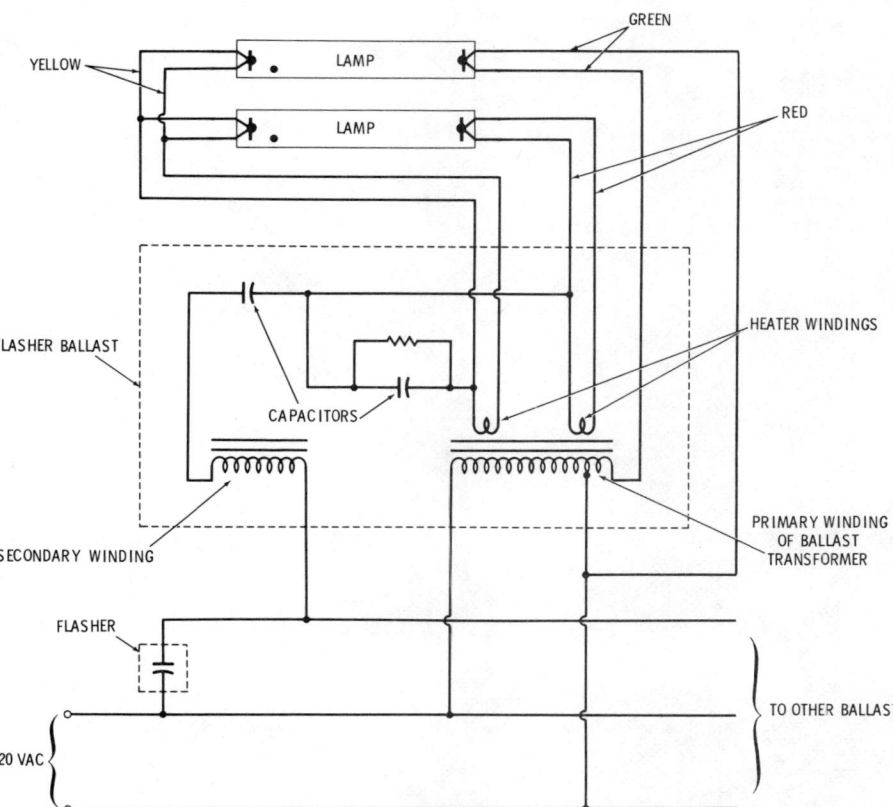

Fig. 25-10. A typical two-lamp flasher-ballast circuit.

tice that this flasher diagram is quite similar to the dimming diagram in Fig. 25-9.

It is also possible to obtain a dimming effect by inserting resistance in place of the flasher. A high dimming ratio cannot be achieved by this method, but interesting effects can be produced by flashing lamps from bright to dim.

SUMMARY

The proper control of lighting plays a very important role in the proper design of lighting systems. An otherwise good lighting design can become a poor design if improperly controlled. We urge the study of manufacturers' catalogs and wiring diagrams for more uses of the items described in this unit. The catalogs are usually free of charge and contain information invaluable to the lighting designer. These may be obtained from your local lighting showroom or by writing directly to the manufacturer; addresses of many may be found in electrical trade journals.

SECTION IX

APPENDIXES

APPENDIX A

Recommended Illumination Levels*

	Recommended Footcandles
Art Galleries	
General	30
On paintings (supplementary)	30
On statuary and other displays	100
Auditoriums	
Assembly	15
Exhibition	30
Banks	
Lobby	
General	50
Writing areas	70
Tellers' stations, posting, and keypunch	150
Building Construction	
General construction	10
Excavation work	2
Building Exteriors and Monuments, Floodlighted	
Bright surroundings	
Light surfaces	15
Dark surfaces	50
Dark surroundings	
Light surfaces	5
Dark surfaces	20
Bulletins and Poster Boards (Exterior) (Water Tanks or Stacks With Advertising Messages, Flags)	
Bright surroundings	
Light surfaces	50
Dark surfaces	100
Dark surroundings	
Light surfaces	20
Dark surfaces	50
Coal Yards (Protective)	0.2
Depots, Terminals, and Stations	
Waiting rooms, rest and smoking rooms	30
Ticket offices: general, ticket racks, counters	100
Baggage checking	50
Platforms and storage	20
Toilets and washrooms	30

	Recommended Footcandles
Dredging	2
Hospitals	
Anesthetizing and preparation room	30
Autopsy and morgue	
Autopsy room	100
Autopsy table	1000
Morgue, general	20
Central sterile supply	
General	30
Needle sharpening	150
Dental suite	
Operatory, general	70
Instrument cabinet	150
Dental chair	1000
Laboratory, bench	100
Recovery room	5
Emergency room	
General	100
Local	2000
Examination and treatment room	
General	50
Examining table	100
Exits, at floor	5
Eye, ear, nose and throat suite	
Darkroom (variable)	0-10
Eye examination room, ear, nose & throat room	50
Fracture room	
General	50
Fracture table	200
Laboratories	
General	50
Close work	100
Libraries	70
Locker rooms	20
Lobby and lounge rooms (lobby daytime 50)	30
Medical records room	100
Nurses' station	
General—Day	70
General—Night	30
Desk and charts	50
Medicine room counter	100

*Courtesy Illuminating Engineering Society

	Recommended Footcandles
Nurses' workroom	30
Nurseries	
General	30
Examination table	100
Play room, pediatric	30
Obstetrical	
Clean-up and scrub-up rooms	30
Labor room	20
Delivery room, general	100
Delivery table	2500
Pharmacy	
General, manufacturing	50
Work table	100
Active storage	30
Private rooms and wards	
General	10
Reading	30
Psychiatric disturbed patients' areas	10
Radioisotope facilities	
Radiochemical laboratory	30
Up-take measuring room	20
Examination table	50
Solariums	20
Storage, central	
General	30
Office	70
Surgery	
Instrument and sterile supply room	30
Cleanup room (instruments)	100
Operating room, general	100
Operating table	2500
Recovery room	30
Therapy	
Physical	20
Occupational	30
Toilets	30
Utility room	20
Waiting room	
General	20
Reading	30
X-ray room and facilities	
Radiography, fluoroscopy and darkroom	10
Deep and superficial therapy	10
Viewing room	30
Filing room, developed films	30
Storage, undeveloped films	10
Hotels	
Bars and cocktail lounges (see restaurants)	
Bathrooms	
General	10
At mirror	30
Bedrooms	
General	10
Makeup	30
Reading, ink writing	30
Dining areas (see restaurants)	
Entrance foyer	30
Front office	50
Laundry	
Washing	30
Flat work ironing	50
Machine and press finishing	70
Linen room	
General	20
Sewing	100
Lobby	
General	10
Reading and working areas	30
Marquee	
Dark surroundings	30
Bright surroundings	50
Loading Platforms	20
Lumber Yards	1
Municipal Buildings; Fire and Police	
Police	
Identification records	150
Jail cells and interrogation rooms	30
Fire hall	
Dormitory	20
Wagon room and recreation room	30
Museums (See Art Galleries)	
Offices	
Reading high-contrast or well-printed material; tasks and areas not involving critical or prolonged seeing such as conferring, interviewing, inactive files, and washrooms	30
Reading or transcribing handwriting in ink or medium pencil on good-quality paper; intermittent filing	70
Regular office work; reading good reproductions, reading or transcribing handwriting in hard pencil or on poor paper; active filing; index references; mail sorting	100
Accounting, auditing, tabulation, bookkeeping, business machine operation; reading poor reproductions; rough layout drafting	150
Cartography, designing, detailed drafting	200
Corridors, elevators, escalators, stairways	20
Parking Lots	5
Self-parking	1
Attendant parking	2
Active shopping centers	2
Piers, Freight and Passenger	20
Post Offices	
Lobby, on tables	30
Sorting, mailing, etc.	100
Prison Yards	5
Quarries	5
Railroad Yards—Classification	
Switch points	2
Body of yard	1
Residences	
Specific visual tasks	
Table games	30
Kitchen activities	
Sink	70
Range and work surfaces	50
Laundry, trays, ironing board, ironer	50

RECOMMENDED ILLUMINATION LEVELS

	Recommended Footcandles
Reading and writing, including studying	
Books, magazines, newspapers	30
Handwriting, reproductions, poor copies	70
Study desks	70
Reading music scores	
Simple scores	30
Advance scores	70
Sewing	
Occasional periods, coarse thread, large stitches, high contrast thread to fabric	30
Occasional periods, light fabrics	50
Prolonged periods, light to medium fabrics	100
Dark fabrics, fine detail, low contrast	200
Shaving, makeup, grooming: on the face at mirror locations	50
Work shop, bench work	70
General lighting	
Entrances, hallways, stairways, stair landings	10
Living room, dining room, bedroom, family room, sun room, library, game or recreation room	10
Kitchen, laundry, bathroom	30
Restaurants, Lunch room, Cafeterias	
Dining areas	
Intimate type	
Subdued environment	3
Light environment	10
For cleaning	20
Leisure type	
Subdued environment	15
Light environment	30
Quick-service type	
Normal surroundings	50
Bright surroundings	10
Cashier	50
Food displays: twice the general levels, but not less than	50
Kitchen	
Inspection, checking, and pricing	70
Other areas	30
Schools	
Reading printed material	30
Reading pencil material	70
Reading spirit-duplicated material	
Good	30
Poor	100
Drafting, bench work	100
Lip reading, chalkboards	150
Sewing	150
Service Stations (at Grade)	
Light surroundings	
Approach	3
Pump island area	30
Service areas	7
Dark surroundings	
Approach	1.5
Pump island area	20
Service areas	3
Shipyards	
General	5
Ways	10
Fabrication area	30
Storage Yards, Active	20

SPORTS LIGHTING

	Recommended Footcandles
Archery (at Shooting Line and Vertical on Target)	
Tournament	10
Recreational	5
Badminton	
Tournament	30
Club	20
Recreational	10

Baseball

	Infield	Outfield
Major league	150	100
AAA and AA league	70	50
A and B league	50	30
C and D league	30	20
Semipro and municipal	20	15
Junior league	30	20
Recreational	15	10
Seats during game		2
Seats before and after game		5

	Recommended Footcandles
Basketball	
College and professional	50
College intramural and high school	30
Recreational (outdoor)	10
Bathing Beaches (Surf)	
On land	1
150 feet from shore	3
Billiards	
Tournament (on table)	50
Recreational (on table)	30
General Area	10
Bowling—(Visual Need Only)	

	Lanes	Pins
Tournament	20	50
Recreational	10	30

	Recommended Footcandles
Bowling on the Green—Same as Croquet	
Boxing or Wrestling	
Championship (ring)	500
Professional (ring)	200
Amateur (ring)	100
Seats during bout	2
Seats before and after bout	5
Croquet	
Tournament	10
Recreational	5
Football (Regulation and Rugby)	
Class I	100
Class II	50
Class III	30
Class IV	20
Class V	10

PRINCIPLES OF ILLUMINATION

	Recommended Footcandles
Golf Driving Range	
General on the tees	10
At 200 yards	5
Golf Putting Greens	10
Gymnasiums—(see Individual Sports)	
Exhibitions and matches	50
General exercise, recreation	30
Locker and shower rooms	20
Handball or Squash	
Tournament	50
Club	30
Recreational	10
Hockey, Field	20
Hockey, Ice—Indoor	
College and professional	100
Amateur league	50
Recreational	20
Horseshoe Pitching—Same as Croquet	
Playgrounds	5
Race Tracks	
Horse, midget auto, motorcycle, bicycle	20
Dog	30
Rifle Range	
On target	100
Firing point	20
Range	10
Roque—Same as Volleyball	
Shuffleboard—Same as Croquet	
Skating	
Rink (indoor—roller or ice)	10
Pond or flooded area	1
Skeet Shooting	
Target surface at 60 feet	30
Firing point, general	10
Ski Slope, Practice	1
Soccer—see Football	

Softball

	In-field	Out-field
Pro and championship	50	30
Semipro	30	20
Industrial league	20	15
Recreational	10	7

	Recommended Footcandles
Swimming Pools	
Underwater, indoor pool	100
Underwater, outdoor pool	60
General, overhead	10

Tennis—Lawn

	Indoor	Outdoor	Table
Tournament	50	30	50
Club	30	20	30
Recreational	20	10	20

	Recommended Footcandles
Trapshooting	
Target surface at 100 feet	30
Firing point, general	10
Volleyball	
Tournament	20
Recreational	10
Stores	
Show windows	
Daytime lighting	
General	200
Feature	1000
Nighttime lighting	
Secondary business districts or small towns	
General	100
Feature	500
Main business districts, highly competitive	
General	200
Feature	1000
Store interiors	
Circulation areas	30
Merchandising areas	
Service	100
Self-service	200
Showcases and wall cases	
Service	200
Self-service	500
Feature displays	
Service	500
Self-service	1000

INDUSTRIAL INTERIORS

	Recommended Footcandles
Airplane Manufacturing	
Stock parts	
Production	100
Inspection	200
Parts manufacturing	
Drilling, riveting, and screw fastening	70
Spray booths	100
Sheet aluminum layout and template work; shaping and smoothing of small parts for fuselage; wing sections, cowling, etc.	100
Subassembly: landing gear, fuselage, wing sections, cowling, and other large units	100
Final assembly and inspection	100
Machine tool repairs	100
Airplane Hangars: Repair Service Only	100
Assembly	
Rough easy seeing	30
Rough difficult seeing	50
Medium	100
Fine	500
Extra fine	1000
Automobile Manufacturing	
Frame assembly	50
Chassis assembly line	100
Final assembly and inspection line	200
Body manufacturing	
Parts	70
Finishing and inspecting	200
Bakeries	
Mixing room	50
Face of shelves (vertical illumination)	30

RECOMMENDED ILLUMINATION LEVELS

	Recommended Footcandles
Inside of mixing bowl (vertical mixers)	50
Fermentation room	30
Makeup room	
Bread	30
Sweet yeast raised products	50
Oven, proofing, and wrapping rooms	30
Fillings and other ingredients	50
Decorating and icing	
Mechanical	50
Hand	100
Book Binding	
Folding, assembling, pasting, etc.	70
Cutting, punching, stitching	70
Embossing and inspection	200
Candy Making	
Chocolate department	
Husking, winnowing, fat extraction, crushing and refining, feeding	50
Bean cleaning and sorting, dipping, packing, wrapping	50
Milling	100
Cream-making: mixing, cooking, and molding	50
Gum drops and jellied forms	50
Hand decorating	100
Hard candy	
Mixing, cooking, and molding	50
Die cutting and sorting	100
Kiss-making and wrapping	100
Canning and Preserving	
Initial grading raw material samples	50
Tomatoes	100
Color grading (cutting rooms)	200
Preparation	
Preliminary sorting	
Apricots and peaches	50
Tomatoes	100
Olives	150
Cutting and pitting, final sorting	100
Canning	
Continuous belt and sink canning	100
Hand packing	50
Olives	100
Examination of canned samples	200
Central Station Indoor Locations	
Auxiliaries, battery rooms, boiler feed pumps, tanks, compressors, and gauge area	20
Boiler platforms, cable room, and circulator or pump bay	10
Burner platform	20
Condensors; deaerator, evaporator, and heater floors	10
Control rooms	
Vertical face of switchboards	
Simplex or section of duplex facing operator	
Type A—Large centralized control room at 66 inches above floor	50
Type B—Ordinary control room at 66 inches above floor	30
Section of duplex facing away from operator	30
Bench boards (horizontal level)	50
Area inside duplex switchboards	10
Rear of all switchboard panels (vertical)	10
Emergency lighting, all areas	3
Chemical laboratory	50
Screen house, switchgear power, telephone equipment room	20
Tunnels or galleries, piping	10
Turbine bay sub-basement	20
Turbine room	30
Chemical works	
Furnaces, driers, stills, evaporators, filtration, bleaching	30
Tanks, crystallizers, extractors, percolators, nitrators	30
Clay Products and Cements	
Grinding, filter presses, kiln rooms	30
Molding, pressing, cleaning, and trimming	30
Color and glazing, rough work; enameling	100
Color and glazing, fine work	300
Cleaning and Pressing Industry	
Checking and sorting	50
Dry and wet cleaning and steaming	50
Inspection and spotting	500
Pressing, machine and hand	150
Repair and alteration	200
Cloth Products	
Cloth inspection	2000
Cutting and pressing	300
Sewing	500
Coal Tipples and Cleaning Plants	
Breaking and cleaning areas	10
Picking	300
Dairy Products: Fluid Milk Industry	
Boiler room and bottle storage	30
Bottle sorting	50
Can washing and cooling equipment	30
Filling: inspection	100
Gauges, meter panels, and thermometers (on face)	50
Laboratories	100
Pasteurizers, separators, and storage refrigerators	30
Tanks, vats	
Light interiors	20
Dark interiors	100
Electrical Equipment Manufacturing	
Impregnating	50
Insulating: coil winding	100
Testing	100
Flour Mills	
Rolling, sifting, purifying	50
Packing	30
Product control	100
Cleaning screens, man lifts, walkways, bin checking	30

PRINCIPLES OF ILLUMINATION

	Recommended Footcandles
Forge Shops	50
Foundries	
Annealing, cleaning, shakeout	30
Core-making, medium	50
Core-making, fine	100
Grinding and chipping	100
Inspection, medium	100
Molding, large; pouring and sorting	50
Molding, medium	100
Cupola	20
Garages: Automobile and Truck	
Service garages	
Repairs	100
Active traffic areas	20
Parking garages	
Entrance	50
Traffic lanes	10
Storage	5
Glass Works	
Mix and furnace rooms, pressing and lehr, glass blowing machines	30
Grinding, cutting glass to size, silvering	50
Fine grinding, polishing, beveling	100
Inspection, etching and decorating	200
Glove Manufacturing	
Pressing and cutting	300
Knitting and sorting	100
Sewing and inspection	500
Hat Manufacturing	
Dyeing, stiffening, braiding, cleaning, and refining	100
Forming, sizing, pouncing, flanging, finishing, and ironing	200
Sewing	500
Inspection	
Ordinary	50
Difficult	100
Highly difficult	200
Very difficult	500
Most difficult	1000
Iron and Steel Manufacturing	
Open hearth	
Charging floor	20
Pouring slide	
Slag pits	20
Control platforms	30
Hot top	30
Hot top storage, checker cellar	10
Mixer building	30
Calcining building, skull cracker	10
Rolling mills	
Blooming, slabbing, hot strip, hot sheet	30
Cold strip, merchant, sheared plate	30
Pipe, rod, tube, wire drawing	50
Tin plate mills; tinning, galvanizing, cold strip rolling	50
Motor room, machine room	30
Inspection	
Blackplate, bloom, billet chipping	100
Tinplate, other bright surfaces	200

	Recommended Footcandles
Laundries	
Washing	30
Flatwork ironing, weighing, listing, and marking	50
Machine and press finishing, sorting	70
Fine hand ironing	100
Leather Manufacturing	
Pressing, winding, and glazing	200
Grading, matching, cutting, scarfing, sewing	300
Machine Shops	
Rough bench and machine work	50
Medium bench and machine work, ordinary automatic machines, rough grinding, medium buffing and polishing	100
Fine bench and machine work, fine automatic machines, medium grinding, fine buffing and polishing	500
Extra fine bench and machine work, fine grinding	1000
Materials Handling	
Wrapping, packing, labeling	50
Picking stock, classifying	30
Loading, trucking	20
Inside truck bodies and freight cars	10
Paint Manufacturing	
General	30
Comparing mix and standard	200
Paint Shops	
Dipping, spraying, rubbing, firing, ordinary hand painting and finishing art	50
Fine hand painting and finishing	100
Extra fine hand painting and finishing (automobile bodies, piano cases, etc.)	300
Paper Box Manufacturing: General Area	50
Paper Manufacturing	
Beaters, grinding, calendaring	30
Finishing, cutting, trimming, paper-making machines	50
Hand counting, wet end of paper machine	70
Paper machine reel, paper inspection and laboratories	100
Rewinder	150
Plating	30
Polishing and Burnishing	100
Printing Industries	
Type foundries	
Machine and hand casting; font assembly sorting	50
Matrix-making, dressing type	100
Printing plants	
Color inspection and appraisal	200
Machine composition, composing room	100
Presses	70
Imposing stones and proofreading	150
Electrotyping	
Molding, finishing, leveling molds, routing, trimming	100

RECOMMENDED ILLUMINATION LEVELS

	Recommended Footcandles
Blocking, tinning, electroplating, washing, backing	50
Photoengraving	
Etching, staging, blocking	50
Routing, finishing, proofing, tint-laying, masking	100
Rubber Tire and Tube Manufacturing	
Stock preparation	
Bandury, plasticating, milling	30
Calendaring	50
Fabric preparation: stock cutting, bead building	50
Tube and thread tubing machines	50
Tire building	
Solid tire	30
Pneumatic tire	50
Curing department: tube curing, casing curing	70
Final inspection: tube, casing	200
Rubber Goods, Mechanical	
Stock preparation	
Banbury, plasticating, milling	30
Calendaring	50
Fabric preparation: stock cutting, hose looms	50
Molded and extruded products, and curing	50
Inspection	200
Service Areas	
Stairways, corridors, elevators	20
Toilets and washrooms	30
Sheet Metal Works	
Presses, shears, stamps, punches, spinning, miscellaneous machines, medium bench work	50
Tin plate inspection, galvanized; scribing	200
Shoe Manufacturing, Leather	
Cutting tables, marking, buttonholing, skiving, sorting, vamping, counting, stitching on dark materials	300
Making and finishing: nailers, sole layers, welt beaters and scarfers, trimmers, welters, lasters, edge setters, sluggers, randers, wheelers, treers, cleaning, spraying, buffing, polishing, embossing	200
Shoe Manufacturing, Rubber	
Washing, coating, mill run compounding	30
Varnishing, vulcanizing, calendaring, upper and sole cutting	50
Sole rolling, lining, making and finishing processes	100
Stone Crushing and Screening	
Belt conveyor tubes, main line shafting spaces, chute rooms inside of bins	10
Primary breaker room, auxiliary breakers under bins	10
Screens	20
Storage Rooms and Warehouses	
Inactive	5
Active	
Rough bulky	10
Medium	20
Fine	50
Sugar Refining	
Grading	50
Color inspection	200
Textile Mills, Cotton	
Opening, mixing, picking	30
Carding, drawing, slubbing, roving, spinning, spooling	50
Beaming and slashing on comb	
Grey goods	50
Denims	150
Inspection	
Grey goods (hand turning)	100
Denims (rapidly moving)	500
Automatic tying-in	150
Drawing-in by hand	200
Weaving	100
Textile Mills, Silk and Synthetics	
Manufacturing: soaking, fugitive tinting, and conditioning or setting of twist	30
Winding, twisting, rewinding, coning, quilling, slashing	
Light thread	50
Dark thread	200
Warping (silk or cotton system) on creel, on running ends, on reel, on beam, on warp at beaming	100
Drawing-in, on heddles or reed	200
Weaving	100
Textile Mills, Woolen and Worsted	
Grading	100
Drawing or spinning (frame or mule): white	250
Drawing or spinning (frame or mule): colored	100
Twisting: white	50
Warping: white, at reed: white	100
Warping: colored	100
Warping at reed: colored	300
Winding: white	30
Winding: colored	50
Weaving: white	100
Weaving: colored	200
Grey goods room	
Burling	150
Sewing	300
Folding	70
Wet finishing: fulling, scouring, crabbing, drying	50
Dyeing	100
Dry finishing	
Napping, conditioning, pressing, folding	70
Shearing	100
Inspecting (perching)	2000
Tobacco Products	
Drying, stripping, general	30
Grading and sorting	200

PRINCIPLES OF ILLUMINATION

	Recommended Footcandles
Welding	
General illumination	50
Precision manual arc welding	1000
Woodworking	
Rough sawing and bench work	30
Sizing, planing, rough sanding, medium machine and bench work, glueing, veneering, cooperage	50
Fine bench and machine work, fine sanding and finishing	100

APPENDIX B

Coefficient of Utilization Tables*

Computing the coefficient of utilization is a time-consuming process. Therefore, it is necessary for lighting designers to have tables that are available for easy reference. The data on the following pages contains coefficients of utilization for various surface reflections and room cavity ratios. By using the category number that is listed with each fixture in conjunction with the charts in Unit 9, the maintenance factor of the lamp can be determined. Another feature of this data is the "Spacing Not to Exceed" column, which lists maximum permissible ratios of spacing to mounting height above the work plane for the lighting fixture.

*Courtesy Westinghouse Electric Corp.

PRINCIPLES OF ILLUMINATION

COEFFICIENTS OF UTILIZATION

LUMINAIRE	DISTRIBUTION	Spacing Not to Exceed	Ceiling Cavity	Reflectances									
				80%			50%			10%			0%
			Walls	50%	30%	10%	50%	30%	10%	50%	30%	10%	0%
			RCR	Coefficients of Utilization								

Category III — Ventilated Dome Reflector
Spacing: 1.3 x Mounting Height · 0/79

RCR	80/50	80/30	80/10	50/50	50/30	50/10	10/50	10/30	10/10	0/0
1	.85	.82	.79	.79	.77	.75	.73	.72	.71	.69
2	.74	.69	.65	.70	.66	.62	.65	.62	.59	.58
3	.65	.60	.54	.62	.57	.53	.57	.54	.51	.49
4	.58	.51	.46	.55	.49	.45	.51	.47	.44	.42
5	.50	.44	.38	.47	.42	.37	.45	.40	.36	.35
6	.44	.38	.33	.43	.36	.32	.40	.35	.32	.30
7	.40	.33	.28	.38	.33	.28	.36	.32	.27	.26
8	.36	.29	.24	.34	.28	.24	.32	.27	.23	.22
9	.33	.25	.20	.31	.25	.20	.29	.24	.20	.18
10	.29	.22	.18	.28	.22	.18	.26	.21	.18	.17

Category I — R-52 Filament Reflector Lamp, Wide Dist.—500- and 750-Watt
Spacing: 1.5 x Mounting Height · 0/100

RCR	80/50	80/30	80/10	50/50	50/30	50/10	10/50	10/30	10/10	0/0
1	1.08	1.05	1.02	1.01	.99	.97	.94	.93	.91	.89
2	.98	.93	.89	.93	.89	.86	.88	.85	.82	.80
3	.89	.83	.78	.85	.80	.76	.80	.76	.73	.71
4	.81	.74	.68	.77	.72	.67	.73	.69	.65	.64
5	.73	.66	.60	.70	.64	.59	.56	.62	.58	.56
6	.67	.59	.53	.64	.58	.52	.61	.56	.52	.50
7	.60	.52	.47	.58	.51	.46	.55	.50	.46	.45
8	.54	.46	.40	.52	.45	.40	.49	.44	.40	.38
9	.48	.40	.35	.46	.39	.35	.44	.38	.34	.33
10	.43	.36	.30	.42	.35	.30	.40	.34	.30	.28

Category I — R-57 Filament Reflector Lamp, Narrow Dist.—500- and 750-Watt
Spacing: .6 x Mounting Height · 0/100

RCR	80/50	80/30	80/10	50/50	50/30	50/10	10/50	10/30	10/10	0/0
1	1.10	1.08	1.05	1.04	1.02	1.00	.97	.96	.95	.93
2	1.02	.98	.94	.97	.94	.91	.91	.89	.88	.86
3	.95	.90	.85	.91	.87	.83	.86	.83	.81	.79
4	.88	.82	.78	.85	.80	.76	.81	.77	.75	.73
5	.82	.76	.71	.79	.74	.70	.76	.72	.69	.67
6	.77	.70	.66	.74	.69	.65	.72	.68	.64	.63
7	.71	.65	.61	.69	.64	.60	.67	.63	.60	.58
8	.66	.60	.56	.65	.59	.56	.63	.58	.55	.54
9	.62	.55	.51	.60	.55	.51	.59	.54	.50	.49
10	.58	.51	.47	.56	.51	.47	.55	.50	.46	.45

Category III — Ventilated Porcelain Enamel Low Bay, 400-W Phos. Coated Vapor Lamp
Spacing: 1.2 x Mounting Height · 0/76

RCR	80/50	80/30	80/10	50/50	50/30	50/10	10/50	10/30	10/10	0/0
1	.81	.78	.76	.76	.74	.72	.71	.69	.68	.67
2	.73	.69	.65	.69	.66	.63	.64	.62	.60	.59
3	.65	.60	.56	.62	.58	.55	.58	.55	.53	.51
4	.59	.53	.49	.56	.52	.48	.53	.50	.47	.45
5	.53	.47	.43	.51	.46	.42	.48	.44	.41	.40
6	.48	.42	.38	.46	.41	.37	.44	.40	.37	.35
7	.43	.37	.33	.41	.36	.32	.39	.36	.32	.31
8	.39	.33	.29	.38	.32	.28	.36	.32	.28	.27
9	.36	.30	.26	.34	.29	.25	.33	.28	.25	.24
10	.32	.27	.23	.31	.26	.23	.30	.25	.22	.21

Category III — 18″ Ventilated Alum. High Bay Conc. Dist. 400-W Clear Vapor Lamp
Spacing: .7 x Mounting Height · 9/77

RCR	80/50	80/30	80/10	50/50	50/30	50/10	10/50	10/30	10/10	0/0
1	.93	.90	.88	.85	.83	.82	.76	.75	.74	.72
2	.86	.82	.79	.79	.77	.74	.72	.70	.69	.67
3	.79	.75	.71	.74	.70	.68	.68	.65	.64	.62
4	.74	.69	.65	.69	.65	.62	.64	.61	.59	.57
5	.68	.63	.59	.64	.60	.57	.60	.57	.54	.53
6	.63	.58	.54	.60	.56	.52	.56	.53	.50	.49
7	.59	.53	.49	.56	.51	.48	.52	.49	.46	.45
8	.55	.49	.45	.52	.47	.44	.49	.45	.43	.41
9	.50	.45	.41	.48	.43	.40	.45	.42	.39	.38
10	.47	.41	.38	.45	.40	.37	.42	.38	.36	.35

Category III — 18″ Ventilated Alum. High Bay Spread Dist. 400-W Coated Vapor Lamp
Spacing: 1.2 x Mounting Height · 10/74

RCR	80/50	80/30	80/10	50/50	50/30	50/10	10/50	10/30	10/10	0/0
1	.88	.86	.84	.80	.79	.77	.71	.70	.69	.67
2	.81	.77	.74	.75	.72	.70	.67	.65	.64	.62
3	.74	.70	.66	.69	.65	.62	.62	.60	.58	.56
4	.68	.63	.59	.64	.60	.57	.58	.55	.53	.51
5	.63	.57	.53	.59	.55	.51	.54	.51	.49	.47
6	.58	.52	.48	.54	.50	.46	.50	.47	.44	.43
7	.53	.47	.43	.50	.45	.42	.46	.43	.40	.39
8	.48	.43	.39	.46	.41	.38	.42	.39	.36	.35
9	.44	.39	.35	.42	.37	.34	.39	.35	.33	.31
10	.41	.35	.31	.39	.34	.30	.36	.32	.39	.28

Category III — 24″ Ventilated Porcelain Enamel 1000-W Phosphor Coated Vapor Lamp
Spacing: 1.3 x Mounting Height · 11/73

RCR	80/50	80/30	80/10	50/50	50/30	50/10	10/50	10/30	10/10	0/0
1	.86	.83	.80	.78	.76	.73	.68	.67	.65	.63
2	.77	.72	.68	.70	.66	.63	.61	.59	.57	.55
3	.68	.62	.57	.62	.58	.54	.55	.52	.49	.47
4	.61	.55	.49	.56	.51	.47	.50	.46	.43	.41
5	.55	.48	.42	.50	.45	.41	.45	.41	.38	.36
6	.49	.42	.37	.45	.39	.35	.40	.36	.33	.31
7	.43	.36	.31	.40	.34	.30	.36	.31	.28	.26
8	.39	.32	.28	.36	.30	.26	.32	.28	.25	.23
9	.35	.28	.24	.33	.27	.23	.29	.25	.22	.20
10	.32	.25	.21	.29	.24	.20	.26	.22	.19	.17

COEFFICIENT OF UTILIZATION TABLES

COEFFICIENTS OF UTILIZATION

LUMINAIRE	DISTRIBUTION	Spacing Not to Exceed	Ceiling Cavity	80%			50%			10%			0%
			Walls	50%	30%	10%	50%	30%	10%	50%	30%	10%	0%
			RCR	Coefficients of Utilization									
Category III — 24" Ventilated Alum. High Bay Spread Dist. 1000-W Phos. Ctd. Vapor Lamp	7↑ / 79↓	1.0 x Mounting Height	1 2 3 4 5 6 7 8 9 10	.91 .83 .75 .68 .61 .55 .50 .45 .41 .37	.88 .78 .69 .62 .55 .49 .43 .39 .34 .31	.86 .75 .65 .57 .50 .44 .38 .34 .30 .27	.84 .77 .70 .63 .57 .52 .47 .43 .39 .35	.82 .73 .65 .58 .52 .47 .41 .37 .33 .30	.80 .71 .62 .55 .48 .43 .37 .33 .29 .26	.75 .70 .64 .58 .53 .48 .43 .39 .36 .33	.74 .67 .61 .55 .49 .44 .39 .35 .32 .28	.73 .66 .58 .52 .46 .41 .36 .32 .28 .25	.71 .64 .56 .50 .44 .39 .34 .30 .27 .24
Category III — 24" Ventilated Alum. High Bay 1000-W Phos. Coated Vapor Lamp	12↑ / 73↓	1.3 x Mounting Height	1 2 3 4 5 6 7 8 9 10	.90 .83 .77 .71 .65 .60 .55 .51 .47 .44	.88 .79 .72 .66 .60 .55 .50 .45 .41 .38	.86 .76 .68 .62 .56 .50 .46 .41 .38 .34	.81 .76 .70 .66 .61 .56 .52 .48 .44 .41	.80 .73 .67 .62 .57 .52 .47 .43 .40 .37	.78 .71 .64 .59 .53 .48 .44 .40 .37 .33	.71 .67 .63 .59 .55 .52 .48 .44 .41 .38	.70 .66 .61 .57 .52 .48 .44 .41 .38 .35	.70 .64 .59 .55 .50 .46 .42 .38 .35 .32	.67 .62 .57 .53 .48 .44 .40 .37 .34 .31
Category III — 2 T-12 Lamps—Any Loading For T-10 Lamps—C.U. x 1.02	10↑ / 75↓	1.3 x Mounting Height	1 2 3 4 5 6 7 8 9 10	.88 .77 .68 .60 .52 .47 .42 .37 .33 .30	.84 .71 .61 .52 .45 .39 .34 .30 .26 .23	.81 .66 .56 .47 .39 .34 .29 .25 .21 .19	.79 .70 .61 .54 .48 .43 .38 .34 .31 .28	.77 .65 .56 .49 .42 .37 .32 .28 .25 .22	.74 .62 .52 .44 .37 .32 .28 .24 .21 .18	.69 .61 .54 .48 .43 .38 .34 .31 .28 .25	.68 .59 .51 .44 .38 .34 .30 .26 .23 .20	.66 .56 .48 .41 .35 .30 .26 .22 .19 .17	.64 .54 .46 .39 .33 .28 .24 .21 .18 .15
Category II — 2 T-12 Lamps—Any Loading For T-10 Lamps—C.U. x 1.02	17↑ / 71↓	1.3 x Mounting Height	1 2 3 4 5 6 7 8 9 10	.88 .77 .68 .60 .53 .47 .42 .38 .34 .31	.85 .71 .61 .53 .45 .39 .34 .30 .26 .24	.81 .67 .56 .47 .40 .34 .29 .25 .22 .19	.77 .68 .60 .53 .47 .42 .38 .34 .30 .26	.75 .64 .55 .48 .41 .36 .31 .28 .24 .22	.73 .60 .51 .43 .36 .31 .27 .23 .20 .18	.65 .57 .51 .45 .40 .36 .32 .29 .26 .24	.64 .55 .48 .42 .36 .31 .28 .24 .21 .19	.62 .53 .45 .38 .33 .28 .24 .21 .18 .16	.59 .50 .42 .36 .30 .26 .22 .19 .16 .14
Category II — 2 T-12 Lamps—Any Loading Center Shield For T-10 Lamps—C.U. x 1.02	18↑ / 66↓	1.3 x Mounting Height	1 2 3 4 5 6 7 8 9 10	.84 .75 .66 .59 .52 .47 .42 .38 .34 .31	.81 .70 .60 .52 .45 .40 .35 .31 .27 .24	.78 .65 .56 .47 .40 .35 .30 .26 .22 .20	.74 .66 .59 .52 .46 .42 .37 .34 .30 .27	.72 .62 .54 .47 .41 .36 .32 .28 .25 .22	.70 .59 .51 .43 .37 .32 .28 .24 .21 .18	.61 .55 .49 .44 .39 .36 .32 .29 .26 .23	.60 .53 .47 .41 .36 .32 .28 .25 .22 .19	.59 .51 .44 .38 .33 .29 .25 .22 .19 .17	.56 .48 .42 .36 .31 .27 .23 .20 .17 .15
Category III — 3 T-12 Lamps—430 or 800 MA For T-10 Lamps—C.U. x 1.02	9↑ / 74↓	1.3 x Mounting Height	1 2 3 4 5 6 7 8 9 10	.86 .75 .67 .59 .52 .46 .41 .37 .33 .30	.83 .70 .60 .52 .45 .39 .34 .30 .26 .23	.80 .66 .55 .47 .39 .34 .29 .25 .22 .19	.78 .69 .61 .54 .48 .43 .38 .34 .31 .28	.76 .65 .56 .49 .42 .37 .32 .28 .25 .22	.73 .61 .52 .44 .38 .32 .28 .24 .21 .18	.69 .61 .54 .48 .43 .38 .34 .31 .28 .25	.67 .58 .51 .45 .39 .34 .30 .26 .23 .21	.66 .56 .48 .41 .35 .30 .26 .23 .20 .17	.64 .54 .46 .39 .33 .28 .25 .21 .18 .16
Category II — 3 T-12 Lamps—430 or 800 MA For T-10 Lamps—C.U. x 1.02	15↑ / 69↓	1.3 x Mounting Height	1 2 3 4 5 6 7 8 9 10	.85 .75 .66 .59 .51 .46 .41 .37 .33 .30	.82 .70 .60 .52 .44 .39 .34 .30 .26 .23	.79 .65 .55 .46 .39 .33 .29 .25 .21 .19	.76 .67 .59 .52 .46 .41 .37 .33 .30 .27	.73 .63 .54 .47 .40 .35 .32 .27 .24 .21	.71 .59 .50 .43 .36 .31 .27 .23 .20 .18	.64 .57 .51 .45 .40 .36 .32 .29 .26 .23	.63 .55 .48 .41 .36 .31 .28 .24 .21 .19	.62 .52 .45 .38 .33 .28 .24 .21 .18 .16	.59 .50 .42 .36 .30 .26 .23 .19 .16 .14

PRINCIPLES OF ILLUMINATION

COEFFICIENTS OF UTILIZATION

LUMINAIRE	DISTRIBUTION	Spacing Not to Exceed	Ceiling Cavity	80%			50%			10%			0%
			Walls	50%	30%	10%	50%	30%	10%	50%	30%	10%	0%
			RCR	Coefficients of Utilization									
Category V — 2 T-12 Lamps—430 MA For 800 MA—C.U. x .96	12↑ / 60↓	1.5 x Mounting Height	1 2 3 4 5 6 7 8 9 10	.70 .60 .52 .46 .40 .36 .32 .29 .26 .23	.66 .54 .46 .39 .33 .29 .25 .22 .19 .17	.63 .50 .41 .34 .28 .24 .21 .18 .15 .13	.62 .53 .46 .41 .36 .32 .29 .26 .23 .21	.59 .49 .41 .36 .30 .26 .23 .20 .18 .16	.57 .46 .38 .32 .26 .22 .19 .17 .14 .12	.52 .45 .39 .35 .31 .27 .25 .22 .20 .18	.51 .42 .36 .31 .27 .23 .21 .18 .16 .14	.49 .40 .33 .28 .24 .20 .17 .15 .13 .11	.47 .37 .31 .26 .22 .18 .16 .13 .11 .10
Category V — 2 T-12 Lamps—430 MA Prismatic Lens 1' Wide— For T-10 Lamps—C.U. x 1.02	0↑ / 59↓	1.2 x Mounting Height	1 2 3 4 5 6 7 8 9 10	.63 .57 .51 .46 .42 .38 .35 .31 .28 .26	.61 .54 .48 .42 .37 .34 .30 .27 .24 .21	.59 .51 .44 .39 .34 .30 .27 .24 .21 .18	.59 .54 .49 .44 .40 .37 .33 .30 .27 .25	.58 .51 .46 .41 .36 .33 .29 .26 .23 .21	.56 .49 .43 .38 .34 .30 .27 .23 .20 .18	.55 .50 .46 .42 .38 .35 .32 .29 .26 .24	.54 .49 .44 .39 .35 .32 .29 .26 .23 .20	.53 .47 .42 .37 .33 .29 .26 .23 .20 .18	.52 .46 .41 .36 .32 .28 .25 .22 .19 .17
Category V — 2 T-12 Lamps—430 MA Prismatic Lens 2' Wide— For T-10 Lamps—C.U. x 1.01	0↑ / 68↓	1.2 x Mounting Height	1 2 3 4 5 6 7 8 9 10	.73 .66 .59 .53 .48 .44 .39 .36 .32 .29	.71 .62 .55 .48 .43 .38 .34 .30 .27 .24	.68 .59 .51 .45 .39 .34 .30 .26 .23 .20	.69 .62 .56 .51 .46 .42 .38 .34 .31 .28	.67 .59 .53 .47 .42 .37 .33 .30 .26 .23	.66 .57 .50 .44 .39 .34 .30 .26 .23 .20	.64 .58 .53 .48 .44 .40 .36 .33 .29 .27	.62 .56 .50 .45 .40 .36 .32 .29 .25 .23	.61 .55 .48 .43 .38 .33 .30 .26 .23 .20	.60 .53 .47 .41 .36 .32 .28 .25 .21 .19
Category V — 4 T-12 Lamps—430 MA Prismatic Lens 2' Wide— For T-10 Lamps—C.U. x 1.02	0↑ / 62↓	1.2 x Mounting Height	1 2 3 4 5 6 7 8 9 10	.66 .60 .54 .49 .44 .40 .36 .32 .29 .27	.64 .56 .50 .44 .39 .35 .31 .28 .24 .22	.62 .53 .46 .41 .35 .31 .28 .24 .21 .19	.62 .56 .51 .46 .42 .38 .35 .31 .28 .26	.61 .54 .48 .43 .38 .34 .30 .27 .24 .23	.59 .52 .45 .40 .35 .31 .27 .24 .21 .19	.58 .53 .48 .44 .40 .36 .33 .30 .27 .25	.57 .51 .46 .41 .37 .33 .30 .26 .23 .21	.56 .49 .44 .39 .34 .31 .27 .24 .21 .18	.55 .48 .43 .38 .33 .29 .26 .23 .20 .17
Category V — 6 T-12 Lamps—430 MA Prismatic Lens 2' Wide— For T-10 Lamps—C.U. x 1.03	0↑ / 56↓	1.2 x Mounting Height	1 2 3 4 5 6 7 8 9 10	.60 .54 .49 .44 .40 .36 .33 .30 .27 .24	.58 .51 .45 .40 .35 .32 .28 .25 .22 .20	.56 .48 .42 .37 .32 .29 .25 .22 .19 .17	.56 .51 .46 .42 .38 .35 .32 .28 .26 .23	.55 .49 .43 .39 .35 .31 .28 .25 .22 .20	.54 .47 .41 .36 .32 .28 .25 .22 .19 .17	.52 .48 .44 .40 .36 .33 .30 .27 .25 .22	.51 .46 .41 .37 .33 .30 .27 .24 .21 .19	.50 .45 .40 .35 .31 .28 .25 .22 .19 .17	.49 .44 .39 .34 .30 .27 .24 .21 .18 .16
Category V — 8 T-12 Lamps—430 MA Prismatic Lens 4' x 4'— For T-10 Lamps—C.U. x 1.02	0↑ / 55↓	1.3 x Mounting Height	1 2 3 4 5 6 7 8 9 10	.59 .53 .48 .43 .39 .35 .32 .29 .26 .24	.57 .50 .44 .39 .35 .31 .28 .25 .22 .20	.55 .47 .41 .36 .31 .28 .25 .22 .19 .17	.55 .50 .45 .41 .37 .34 .31 .28 .25 .23	.54 .48 .42 .38 .34 .30 .27 .24 .21 .19	.52 .46 .40 .35 .31 .28 .25 .22 .19 .17	.51 .47 .43 .39 .35 .32 .29 .27 .24 .22	.50 .45 .40 .36 .32 .29 .26 .24 .21 .19	.49 .44 .39 .34 .30 .27 .24 .21 .19 .17	.48 .43 .38 .33 .29 .26 .23 .20 .18 .16
Category V — 4 T-12 Lamps—430 MA Prismatic Lens 2' Wide— For T-10 Lamps—C.U. x 1.02	2↑ / 51↓	1.2 x Mounting Height	1 2 3 4 5 6 7 8 9 10	.56 .50 .45 .41 .37 .33 .30 .27 .25 .22	.54 .47 .41 .37 .32 .29 .26 .23 .20 .18	.52 .45 .38 .34 .29 .26 .23 .20 .18 .16	.52 .47 .42 .38 .34 .31 .29 .26 .23 .21	.50 .44 .39 .35 .31 .28 .25 .22 .20 .18	.49 .42 .37 .32 .28 .25 .22 .20 .17 .15	.47 .43 .39 .35 .32 .29 .27 .24 .22 .20	.46 .41 .37 .33 .29 .27 .24 .21 .19 .17	.45 .40 .35 .31 .27 .24 .22 .19 .17 .15	.44 .39 .34 .30 .26 .23 .20 .18 .16 .14

COEFFICIENT OF UTILIZATION TABLES

COEFFICIENTS OF UTILIZATION

LUMINAIRE	DISTRIBUTION	Spacing Not to Exceed	Ceiling Cavity	Reflectances								
				80%			70%			50%		
			Walls	50%	30%	10%	50%	30%	10%	50%	30%	10%
			RCR	Coefficients of Utilization								
Category V — 2 T-12 Lamps—430 MA 1' Wide Prismatic Wrap-Around	7↑ ↓59	1.2 x Mounting Height	1	.68	.65	.63	.65	.63	.61	.61	.60	.58
			2	.60	.56	.53	.58	.55	.52	.55	.52	.49
			3	.54	.49	.45	.52	.48	.45	.50	.46	.43
			4	.49	.43	.40	.47	.43	.39	.45	.41	.38
			5	.44	.38	.34	43	.38	.34	.40	.36	.33
			6	.40	.34	.30	.39	.34	.30	.37	.32	.29
			7	.36	.31	.27	.35	.30	.26	.33	.29	.26
			8	.32	.27	.24	.32	.27	.23	.30	.26	.23
			9	.29	.24	.21	.29	.24	.20	.27	.23	.20
			10	.27	.22	.18	.26	.21	.18	.25	.21	.18
Category V — 4 T-12 Lamps—430 MA 2' Wide Prismatic Wrap-Around	4↑ ↓59	1.3 x Mounting Height	1	.66	.64	.61	.64	.62	.60	.61	.59	.57
			2	.59	.55	.52	.57	.54	.51	.55	.52	.49
			3	.53	.48	.45	.52	.48	.44	.49	.46	.43
			4	.48	.43	.39	.47	.42	.39	.45	.41	.38
			5	.43	.38	.34	.42	.37	.34	.40	.36	.33
			6	.39	.34	.30	.38	.34	.30	.36	.32	.29
			7	.35	.30	.26	.34	.30	.26	.33	.29	.26
			8	.32	.27	.23	.31	.26	.23	.30	.26	.23
			9	.28	.24	.20	.28	.23	.20	.27	.23	.20
			10	.26	.21	.18	.25	.21	.18	.25	.20	.17
Category I — 2 Lamp Strip—Any Loading	17↑ ↓69	1.6 x Mounting Height	1	.83	.79	.75	.79	.76	.72	.73	.70	.67
			2	.71	.65	.60	.68	.62	.57	.62	.58	.54
			3	.62	.55	.50	.59	.53	.47	.55	.49	.44
			4	.55	.47	.41	.52	45	.39	.48	.42	.37
			5	.48	.40	.34	.46	.38	.33	.42	.36	.31
			6	.43	.35	.29	.41	.33	.28	.38	.31	.26
			7	.38	.30	.25	.36	.29	.24	.34	.27	.23
			8	.34	.26	.21	.33	.25	.21	.30	.24	.19
			9	.30	.23	.18	.30	.23	.18	.27	.21	.17
			10	.28	.21	.16	.27	.20	.15	.25	.19	.15
Category V — 1 Lamp—Any Loading 2' Wide, 1' Deep Prismatic Lens	0↑ ↓60	1.2 x Mounting Height	1	.64	.62	.60	.63	.61	.59	.60	.59	.57
			2	.58	.55	.52	.57	.54	.51	.55	.52	.50
			3	.52	.48	.45	.51	.47	.44	.49	.46	.44
			4	.47	.42	.39	.46	.42	.39	.45	.41	.38
			5	.42	.37	.34	.42	.37	.34	.40	.36	.34
			6	.38	.33	.30	.38	.33	.30	.37	.32	.30
			7	.35	.30	.26	.34	.30	.26	.33	.29	.26
			8	.31	.26	.23	.31	.26	.23	.30	.26	.23
			9	.28	.23	.20	.28	.23	.20	.27	.23	.20
			10	.26	.21	.18	.25	.21	.18	.25	.21	.18
Category VI — 2 Lamp—Any Loading Opaque Sides	75↑ ↓5	1.5 x Mounting Height	1	.68	.65	.62	.59	.56	.54	.42	.41	.39
			2	.59	.54	.51	.51	.48	.44	.37	.35	.32
			3	.52	.46	.42	.45	.40	.37	.32	.29	.27
			4	.46	.40	.35	.40	.35	.31	.28	.25	.23
			5	.40	.34	.30	.35	.30	.26	.25	.22	.20
			6	.36	.30	.26	.31	.27	.23	.22	.20	.17
			7	.32	.26	.22	.28	.23	.19	.20	.17	.14
			8	.29	.23	.19	.25	.20	.17	.18	.15	.13
			9	.26	.20	.17	.23	.18	.15	.17	.13	.11
			10	.24	.18	.15	.21	.16	.13	.15	.12	.10
Category VI — Luminous Ceiling—50% Transmission 80% Cavity Reflectance	0↑ ↓69	1.5 to 2.0 x Mounting Height above Diffuser	1	① For cavities that are painted white use 70% effective ceiling cavity reflectance. ② For cavities that are obstructed or have lower reflectances use 50% effective ceiling cavity reflectance.			.60	.58	.56	.58	.56	.54
			2				.53	.49	.45	.51	.47	.43
			3				.47	.42	.37	.45	.41	.36
			4				.41	.36	.32	.39	.35	.31
			5				.37	.31	.27	.35	.30	.26
			6				.33	.27	.23	.31	.26	.23
			7				.29	.24	.20	.28	.23	.20
			8				.26	.21	.18	.25	.20	.17
			9				.23	.19	.15	.23	.18	.15
			10				.21	.17	.13	.21	.16	.13
Category VI — Cove Without Reflector	Cove 12 to 18 inches below ceiling. Reflectors with fluorescent lamps increase coefficients of utilization 5 to 10%.		1	.42	.40	.39	.36	.35	.33	.25	.24	.23
			2	.37	.34	.32	.32	.29	.27	.22	.20	.19
			3	.32	.29	.26	.28	.25	.23	.19	.17	.16
			4	.29	.25	.22	.25	.22	.19	.17	.15	.13
			5	.25	.21	.18	.22	.19	.16	.15	.13	.11
			6	.23	.19	.16	.20	.16	.14	.14	.12	.10
			7	.20	.17	.14	.17	.14	.12	.12	.10	.09
			8	.18	.15	.12	.16	.13	.10	.11	.09	.08
			9	.17	.13	.10	.15	.11	.09	.10	.08	.07
			10	.15	.12	.09	.13	.10	.08	.09	.07	.06

Index

A

Absorption, 29
Adjusted footcandles, 55-56
Advantages of artificial lighting, 133
Air-duct installation, 128
All-weather fluorescent lamps, 41
Alternate speller, 151
Alumina, Polycrystalline, 25
Amber lamps, 108
American National Standards Institute, 35
Angle
    of sight, 14
    visual, 14
Apparel, wearing, 75
Application of electricity for light, 9
Appraisal, 69
Arcs, electric, 9
Area lighting, industrial, 85
Argon, 31, 39
    filling gas, 31
Artificial
    lighting, advantages of, 133
    white light, 17
Atmosphere, 69
Atomic structure, 24
Attraction, 69
Autotransformers, 152

B

Balanced store lighting, 76
Ballast, 41
Base, 29
Baseball fields, lighting, 100
Bathroom lighting, 81
Beam utilization, calculating, 94
Beauty-tone lamps, 28
Bedroom and hall lighting, 81
Billboard lighting, 121
Black light, 123
Blue lamps, 108
Boarder-chaser, 151
Brightness ratios, 63
Bulb, 27
    finish and color, 28
    shapes, 36
    size scale, 28

C

Calculating
    beam-lumen method, 93
    beam utilization, 94
    floodlighting, 93
Candlepower
    distribution, 58
        curves, 57
    of standard lamps, table, 108
Carbon electrodes, 9
Cavity, 45
    ceiling, 49
    floor, 49
    ratio, 49-51
        table, 52
    reflectance table, 53
    room, 49
Ceiling cavity, 49
Centennial exposition, Chicago, 24
Chart, eye test, 14
Chicago centennial exposition, 24
Classes of fluorescent lamps, 40
    instant start, 40
    preheat, 40
    rapid-start, 41
Coefficient of utilization, 51, 95
    table, 53, 165
Coiled-coil filament, 31
Cold-cathode fluorescent lamps, 41
Collector street lighting, 113
Color
    consideration, 70
    industrial lighting, 85
    of bulb, 28
Colored lens caps, 77
Colors
    of fluorescent lamps, 39
        cool white, 39
        daylight, 40
        deluxe cool white, 39
        deluxe warm white, 39
        warm white, 39
        white, 39
    three basic, 17
Combination switch, 148
Common materials, light reflected from, 19
Comparison of lighting cost, 142

Compounds, iodine, 25
Contrast, 14-15
Contribution to twentieth-century progress, 10
Control panel lighting, 25
Cornice lighting, 79
Correction factor, 51
Cost, industrial lighting, 85
Cost-of-lighting comparison table, 141-142
Counter area lighting, 70
Covering, 13
Curves
    candlepower distribution, 57-58
    maintenance factor, 55

D

Daylight, 17
Design, lighting, 64
Devices, dimming, 77
Diagram, polar, 57
Diaphragm, 13
Diffused light, 70
Diffuser, 64
Dimmer, 150-151
Dimming, 42
    devices, 77
    of fluorescent lamps, 151
    room lighting, 80
Distribution
    curves, candlepower, 57
    photometric, 58
Double-coiled filament, 31
Downlights, recessed, 79
Drag strip lighting, 105
Driving-range lighting, 105
Dungeness lighthouse, 9
Dysprosium, 25

E

Edison, Thomas A., 9
Effective reflectance, 51
Efficiency, 37
Electric
    arcs, 9
    discharge lamp, 24-25
    illumination, 10
    lamp today, 9
Electricity, application of for light, 9
Electrodes, 35
    carbon, 9
Electroluminescence, 25
Electroluminescent
    lamps, 23-25
    night light, 25
Emphasis of merchandise, lighting for, 70
Energy, ultraviolet, 37
Entry hall lighting, 79
Envelope, lamp, 23
Equipment, explosive proof, 90
Explosive-proof equipment, 90
Exposition, Chicago centennial, 24
Exposure-time table, 130

Expressway lighting, 115
Extinguished, 37
Eye test chart, 14
Eyelid, 13

F

Factors of good lighting, 83
Family room lighting, 81
Feature display lighting, 75
Field measurements, 19
Filament, 29-30
    coiled coil, 31
    double coiled, 31
Filling gas, 31
Film, 13
Finish, of bulb, 28
Fixture, 46, 64
Flasher
    off-on, 153
    speller, 153
Flashing, 42, 123
    fluorescent lamps, 153
    of lamps, 151
Floodlight, 31-32
Floodlighting calculations, 93
Floor cavity, 49
Fluorescent
    lamp, 18, 24
        data table, 42
        dimming of, 151
        flashing, 153
        use of, 69
    sunlamp, 129
Football field lighting, 101
Footcandle, 18-19
    adjusted, 55-56
    meter, 19-20
Form, lighting survey, 46
Fountain lighting, 107
Four-way switch, 148
Foyer lighting, 79
Freeway lighting, 115
Frequencies, wave, 17

G

Gas-filled lamp, 27
Gaseous discharge lamps, 23
General lighting service lamps, 31
Germicidal lamp types, 127
Glare, 63-64
    reflected, 83
Good lighting, factors of, 83
Green lamps, 108

H

Halogen lamps, 32
High- and low-voltage lamps, 31
High-pressure sodium lamps, 38

INDEX

Highway
  interchange lighting, 119
  lighting, 115
Horizontal plane, 57
Horse race track, 102
Hot-cathode lamp, 40
Housing, 13
How we see, 13
Hydrogen, 31

Illumination
  electric, 10
  levels, 64, 69, 95, 157
  of fluorescent lamps, table, 75
  quality of, 63
  road sign, 25
Images, reflected, 77
Incandescent, 17
  lamps, 9-10, 23, 27
Indium, 25
Indoor sports, lighting, 99
Industrial lighting
  areas, 85
  color, 85
  cost, 85
Infrared lamps, 32-33
Inside-frosted lamp, 18
Installation, in air duct, 128
Interchange lighting, highway, 119
Interpolation, 51, 54
Inverse square law for light, 19
Iodine
  compounds, 25
  vapor, 23
Iris, 13

J

Jousting, lighting for, 136

K

Kitchen lighting, 81
Krypton, 31
  filling gas, 31

L

Lamp(s)
  amber, 108
  and lighting fixtures, 54
  beauty tone, 28
  blue, 108
  electric discharge, 25
  electroluminescent, 23, 25
  envelope, 23
  flashing of, 151
  floodlight, 32
  fluorescent, 18, 24, 39-42
  gas filled, 27
  gaseous discharge, 23

Lamp(s)—cont
  green, 108
  halogen, 32
  high and low voltage, 31
  hot cathode, 40
  incandescent, 9, 23, 27
  infrared, 32
  inside frosted, 18
  instant start, 40
  lumiline, 32
  mercury, 24, 35
  metal halide, 25, 37
  preheat, 40
  projection, 32
  projector, 32, 70
  quartz iodine, 23
  rapid start, 41
  red, 108
  reflector, 32, 70
  rs reflector, 129
  service, 31
  showcase, 32
  size, 17
  sodium, 38
    vapor, 25
  tubing, 24
  tungsten halogen, 23
  turquoise, 108
  types, 36
    germicidal, 127
Lens, 13
  caps, 77
Level of illumination, 95
Light
  beam, 13
  colors, 17
  diffused, 70
  reflection, 19
  source selector table, 96
  sources, choice of, 69
  spectrum, visible, 17
  ultraviolet, 25
Lighthouse, Dungeness, 9
Lighting
  baseball fields, 100
  billboard, 121
  collector street, 113
  control panel, 25
  counter areas, 70
  design, 64
  drag strip, 105
  driving range, 105
  expressway, 115
  feature display, 75
  football fields, 101
  for emphasis of merchandise, 70
  for jousting, 136
  fountain, 107
  freeway, 115
  highway, 115
    interchange, 119
  indoor sports, 99
  industrial area, 85

Lighting—cont
  local and minor street, 113
  major streets, 111
  miniature golf, 105
  mirrors, 75
  outdoor sports, 100
  residential-outlying area, 113
  roadway, 111
  sales areas, 75
  show window, 77
  showcases, 70
  softball fields, 101
  supplementary, 88
  survey, 45-46
  swimming pool, 109
  tennis court, 105
  underpass, 119
  wall cases, 75
  with sunlight, 135
Lighting fixtures, 64
  layout of, 55
  location of, 122
  selector table, 84-85
Living room lighting, 79
Local and minor street lighting, 113
Long life, 36
Lumen, 18
Lumiline lamps, 32
Luminance, 14-15
Luminous wall panels, 79

M

Magnetic relay, 150
Maintenance factor, 54, 95
  curves, 55
  table, 95
Major street lighting, 111
Mandrel, 31
Measurements
  field, 19
  units of light, 17
Mechanism, seeing, 13
Mercury lamps, 24
  designation of, 35
  operating characteristics, 36-37
Mercury switch, 147
Metal-halide lamps, 37
Meter, footcandle, 19-20
Miniature golf area lighting, 105
Mirror lighting, 75

N

Natural Chimneys, 137
Neon, 39
Newton, Sir Isaac, 17
Night light, electroluminescent, 25
Nitrogen, 31
  filling gas, 31
Normal operation, 37
North light, 40

O

Objective factors, process of seeing, 13
Off-on flasher, 153
Operating
  characteristics of mercury lamps, 36-37
  cost of lighting, 142
Operation, normal, 37
Outdoor lighting, 81
  sports, 100
Owning cost of lighting, 141
Oxidation, 35
  of filament, 27

P

Pearl Street station, 9
Phosphor, 24, 25, 39
Photoelectric switch, 149
Photometric distribution, 58
Photoswitch, 149
Photosynthetic and photoperiodic lighting table, 134
Polar diagram, 57
Polycrystalline alumina, 25
Prism, 17
Projection, 31
  lamps, 32
Projector lamps, 32, 70

Q

Quality of illumination, 63
Quartz-iodine lamp, 23
Quiet switch, 148

R

Race track lighting
  drag strip, 105
  horse, 102
  stock car, 104
Radiation, ultraviolet, 37, 39
Rays, 35
Recessed downlights, 79
Recommended
  illumination levels, 157
  reflectances, 64
Red lamps, 108
Reflectance, 63
  ceiling, 51
  effective, 51
  floor, 51
  recommended, 64
  surface, 45
  table, 64
Reflected
  glare, 83
  images, 77
Reflector lamps, 32, 70
Relay, 149-150
Remote positioners, 152
Residential lighting
  bathroom, 81

INDEX

Residential lighting—cont
  bedroom and halls, 81
  dining room, 80
  entry hall, 79
  family room, 81
  kitchen, 81
  living room, 79
  outdoor, 81
  outlying area, 113
Restarting, 37
Retina, 13
Rheostat dimmer, 151
Road sign illumination, 25
Roadway
  illumination table, 111
  lighting, 111
    configurations table, 117
Room
  cavity, 49
  dimension, 45
RS reflector lamp, 129

S

Sales area lighting, 75
Searchlight, 31
Seeing mechanism, 13
Shadows, 63, 83
Shape, bulb, 27
Show window lighting, 77
Showcase
  lamps, 32
  lighting, 70
Shutter, 13
Sight, angle of, 14
Single-pole switch, 147
Size, 14
  bulb, 27
  lamp, 17
  scale, bulb, 28
Snap-action switch, 147
Sodium, 25
  lamps, high pressure, 38
  vapor lamps, 25
Softball field lighting, 101
Speller flasher, 153
Spotlight, 31, 32
Starting, 36
Store lighting, balanced, 76
Structure, atomic, 24
Sunlamp
  fluorescent, 129
  types, 129
Sunlight, lighting with, 135
Supplementary lighting, 88
Surface reflectance, 45
Survey, lighting, 45
Swimming pool lighting, 109
Switch
  combination, 148
  four-way, 148
  mercury, 147
  Photoelectric, 149

Switch—cont
  quiet, 148
  snap action, 147
  three way, 148

T

Table
  candlepower of standard lamps, 108
  cavity ratios, 52
    reflectance, 53
  coefficient of utilization, 53, 165
  cost-of-lighting comparison, 141-142
  exposure time, 130
  fluorescent lamp data, 42
  illumination of fluorescent lamps, 75
  light source selector, 96
  lighting fixture selector, 84-85
  maintenance factor, 95
  Photosynthetic and Photoperiodic lighting, 134
  reflectance in office and school, 64
  roadway illumination, 111
    lighting configurations, 117
  trigonometric values, 59
  watts per square foot, 96
Tennis court lighting, 105
Test chart, eye, 14
Thallium, 25
Three
  basic colors, 17
  way switch, 148
Total cost of lighting, 142
Transformer, 36
Traveler wire, 148
Trigonometric values table, 59
Tubing, lamp, 24
Tungsten, 31
  halogen lamps, 23
Turquoise lamps, 108
Twinkler, 151
Type
  floodlight fixture, 95
  of lamps, 31
  lamps used, 95

U

Ultraviolet
  energy, 37
  light, 25
  radiation, 37, 39
Underpass lighting, 119
Undervoltage, 36
Uniformity, 63
Units of light measurement, 17
Utilization, coefficient of, 51

V

Valance lighting, 79
Vapor, iodine, 23
Vaporize, 37
Vaporizing, of filament, 27

Vertical plane, 57
Visible light spectrum, 17
Visual angle, 14
Voltage, 36

W

Wall
   case lighting, 75
   panels, luminous, 79
Wallwashers, 79
Warm-up, 37

Water disinfection, 128
Watts, 17
   per square foot table, 96
Wave frequencies, 17
Wearing apparel, 75
White light, 17
Working plane, 19-20, 46

Z

Zonal cavity, 45, 49